Ilja Nieuwland
The Lost Termini of Berlin

Ilja Nieuwland

The Lost Termini of Berlin

A Biography of the Railway Stations that Shaped
the German Capital

DE GRUYTER
OLDENBOURG

ISBN 978-3-11-138121-3
e-ISBN (PDF) 978-3-11-138187-9
e-ISBN (EPUB) 978-3-11-138198-5

Library of Congress Control Number: 2024939249

Bibliographic information published by the Deutsche Nationalbibliothek
The Deutsche Nationalbibliothek lists this publication in the Deutsche Nationalbibliografie;
detailed bibliographic data are available on the internet at http://dnb.dnb.de.

For Marieke, Ger, and Mark

Foreword

When I came to live in Berlin in the mid-1990s, one of the many, many things that amazed me was the sheer scale of the city's public transport infrastructure. Although part of it was still in use, even more appeared to serve little purpose. There were endless numbers of railway bridges crossing the Yorckstraße with hardly a train using them, kilometers of overgrown tracks and vast stretches ending in what seemed like the middle of nowhere.

Walking through the woods near my home in the southwestern suburbs, I stumbled upon tracks that clearly hadn't been used in decades. Research revealed that it was not just some random bit of railway but a part of Prussia's very first line from Potsdam to the Potsdamer Bahnhof, Berlin's first railway station, a station that even then was being recreated, as part of the reconstruction of Potsdamer Platz after German reunification. A deeper dive into the history of that station, and the city around it, caused a fascination that never quite went away.

This book is the result of that nearly thirty-year interest in Berlin's stations: their architecture, their technology but perhaps most of all the way in which they related to the city and people that used them. In a way, the same book could be written about other cities, such as London or Paris. The big difference is that the city that I am describing no longer exists. Sure, it is still there, but in a radically different form from what it once was.

However, that was to be expected. If Berlin is known for one thing, it is the transformations it went through as the states of which it was the capital came into being, grew and were toppled, sometimes violently. Today, the traces of Brandenburg, Prussia, the German Empire, Nazi Germany and the GDR are still everywhere to be seen, including in the German capital's railway infrastructure. What I have written is not railway history in the usual sense. Of course it concerns the history of Berlin's transport infrastructure, but it is even more about the buildings themselves and the people that used them to travel away from the capital, sometimes permanently. Or to arrive in the largest German metropolis as I did myself, thirty years ago.

This project would never have been finished had it not been for the help of a few kind people. None have offered me more support, both practical and spiritual, than Ger Dijkstra and Mark Thomas. Mark kindly drew up several excellent maps for this book, and patiently went through my chapters, making sure they contained acceptable and understandable English. Ger's omniscience of rolling stuff on steel bars, and his enthusiasm to share and discuss it, was a huge help, as was his willingness to fact-check my chapters; he also kindly allowed for the use of some of his photographs. Nikolaus Bernau was useful in providing the latest information about the various Berlin museums that are relevant to the story

https://doi.org/10.1515/9783111381879-202

while Milo van de Pol prepared a drawing for the book. My fellow author Matthijs de Ridder helped me work through various dilemmas of phrasing and composition of the story. Finally, I want to thank the lurkers of the r/berlin subreddit, whom I abused as proofreaders for some of the early chapters. This book would have not been possible without either of you.

I was lucky enough to receive assistance from several institutions. My colleagues at the Huygens Institute (Royal Netherlands Academy of Arts and Sciences) proved to be grateful sparring partners at various stages of this book's writing. I wish to thank the German Museum of Technology in Berlin, and particularly Antje Stritzke, for their willingness to give me access to its extensive collections, providing assistance in navigating its somewhat labyrinthine catalogue, and allowing so much of their images to be used. Matthias Wein of the Stadtmuseum Berlin was similarly helpful and patient in (always quickly!) replying to my queries. Finally, the Central City Library of Berlin (Zentrale Landesbibliothek) proved to be an essential repository of sources on this subject in a time when Covid measures sometimes turned accessing literature into a challenge.

Of course, the most important party to thank is still my partner Marieke van der Duin, whose equanimity in the face of long writing days, infuriating procrastination and bouts of creative self-pity shaped the rock on which all of my work rests.

Contents

Maps

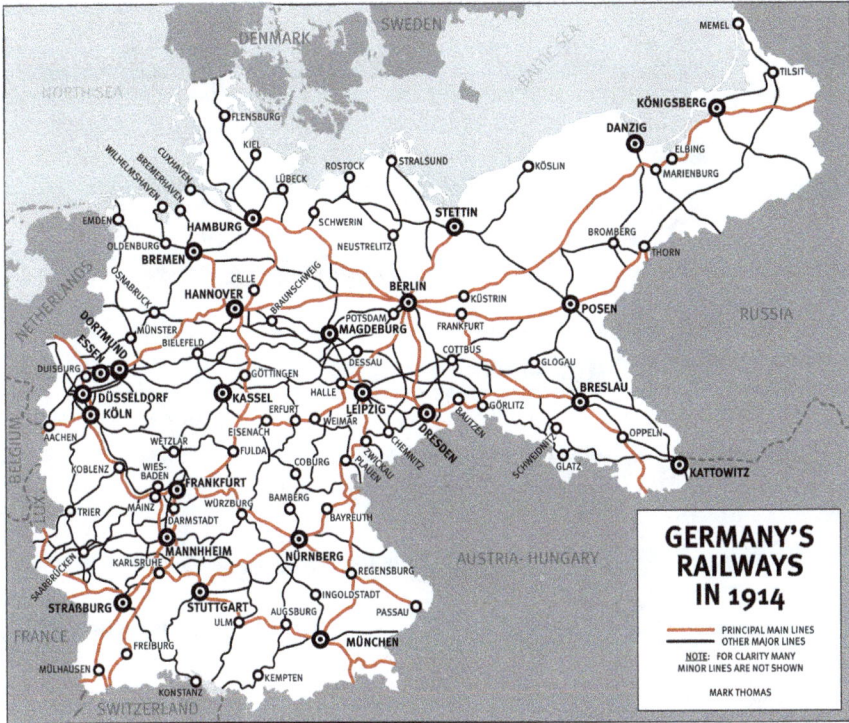

Map 1: Railways in Germany, 1914. Map by Mark Thomas.

https://doi.org/10.1515/9783111381879-204

NOR

WEDDING

SPANDAU,
HANNOVER
& HAMBURG

SPANDAU

JUNGFERNHEIDE

BEUSSELSTR.

PUTLITZSTR.

H

CHARLOTTENBURG

WESTEND

BELLEVUE

LEHRTER
BAHNHOF
(1870)

TIERGARTEN

FRIED

CHARLOTTENBURG

ZOOLOGISCHER
GARTEN

POTSDAMER
BAHNHOF
(1838)

SAVIGNYPLATZ

WANNSEEBAHNHOF

EICHKAMP

RINGBAHNH

HALENSEE

GRUNEWALD

2.

POTSDAM &
MAGDEBURG

SCHMARGENDORF

GROSSGÖRSCHENSTR.

4. 3.

1. SC
2. YO
1.   3. SC
(M
4. DRI
BA
(

WILMERSDORF
- FRIEDENAU

EBERSSTR.

TEM

FRIEDENAU

PAPESTR.

POTSDAM &
MAGDEBURG

SÜDENDE

MARIENDORF

HALLE &
LEIPZIG

DRESDEN

Map 2: Diagram of Berlin's Railways up to 1900. Map by Mark Thomas.

STRALSUND

STETTIN

RUNNEN

SCHÖNHAUSER ALLEE

PRENZLAUER ALLEE

NORDBAHNHOF
(1892)

TINER
NHOF
842)

WEISSENSEE

**DIAGRAM OF
BERLIN'S RAILWAYS
UP TO 1900**

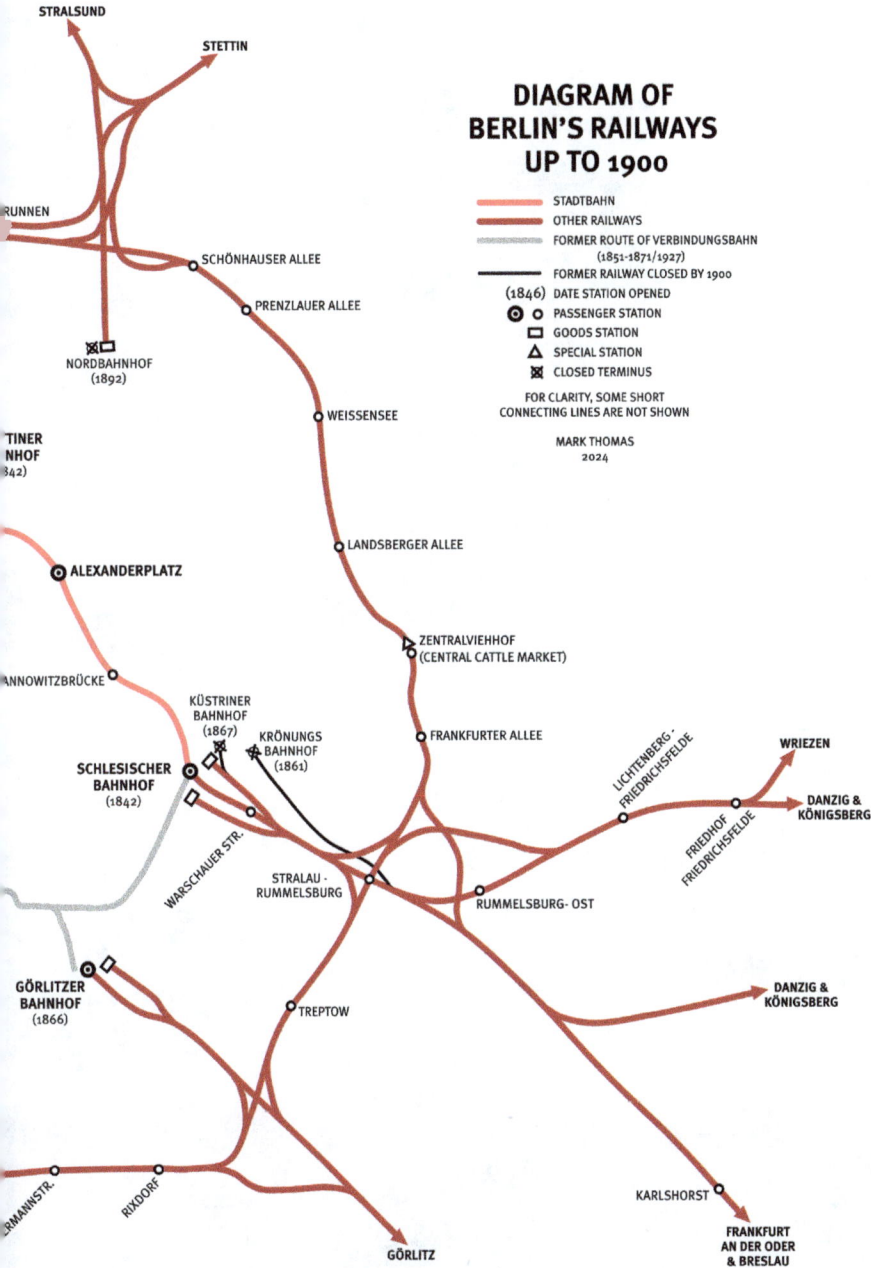

STADTBAHN
OTHER RAILWAYS
FORMER ROUTE OF VERBINDUNGSBAHN
(1851-1871/1927)
FORMER RAILWAY CLOSED BY 1900
(1846) DATE STATION OPENED
PASSENGER STATION
GOODS STATION
SPECIAL STATION
CLOSED TERMINUS

FOR CLARITY, SOME SHORT
CONNECTING LINES ARE NOT SHOWN

MARK THOMAS
2024

LANDSBERGER ALLEE

ALEXANDERPLATZ

ZENTRALVIEHHOF
(CENTRAL CATTLE MARKET)

ANNOWITZBRÜCKE

KÜSTRINER
BAHNHOF
(1867)

KRÖNUNGS
BAHNHOF
(1861)

FRANKFURTER ALLEE

LICHTENBERG ·
FRIEDRICHSFELDE

WRIEZEN

**SCHLESISCHER
BAHNHOF**
(1842)

WARSCHAUER STR.

STRALAU ·
RUMMELSBURG

RUMMELSBURG- OST

FRIEDHOF
FRIEDRICHSFELDE

DANZIG &
KÖNIGSBERG

**GÖRLITZER
BAHNHOF**
(1866)

TREPTOW

DANZIG &
KÖNIGSBERG

ERMANNSTR.

RIXDORF

KARLSHORST

GÖRLITZ

FRANKFURT
AN DER ODER
& BRESLAU

1 Introduction: A City to Arrive in

A Scene at Stettin Station

Fig. 1.1: Stettiner Bahnhof, façade, c. 1930. Colorized Postcard.

A taxi drives up Invalidenstraße, slowly pushes its way through a confusion of pedestrians and electric trams, reaches the square in front of the station, and, honking its horn as if relieved, hurries across the driveway to Stettin station. It stops.

– A lady gets out. "How much?" she asks the driver.

– "Two sixty, *meine Dame*."

The lady was already rummaging in her purse, but now she withdraws her hand. "Two sixty for a ten-minute ride? Nah, dear man, I'm not a millionaire, let my son pay for it. Wait."

– "No can do, lady," says the driver.

– "What do you mean, can't? I won't pay it, so you'll have to wait till my son comes. Four ten on the train from Stettin."

– "Can't. We're not allowed to stop here in the driveway."

– "Then wait over there, *Männeken*. We'll come over, we'll get in over there."

https://doi.org/10.1515/9783111381879-001

The driver puts his head to one side and blinks at the lady. "They're coming, lady," he says. "They're coming as sure as the next pay cut is coming. But you know, let your son give you the money back. Much easier for you, ain't it?"
– "How's this?" asks a *Schupo*.[1] "Drive on, chauffeur."
– "The lady wants me to wait, *Herr Hauptwachtmeister*."[2]
– "Drive on, chauffeur."
– "She won't pay!"
– "Pay please, lady. We can't do that here, other people want to leave too."
–"I don't want to. I'll be right back."
–"I want my money, you old painted . . ."
– "I'll write you up, chauffeur!"

Another driver pipes up. "Drive on, you old goofball, or I'll run you into your Bugatti . . ."
– "So, *gnädige Frau*, please pay! You can see for yourself . . ." In desperation, the Schupo engages in a kind of dancing lesson bow, heels clicking.

The lady beams. "But of course I'll pay. If the man can't wait, I don't want anything to happen that is forbidden. All this excitement! God, Mr. Schupo, we women should regulate such things. Everything would run just so much smoother . . ." – *Hans Fallada, Kleiner Mann was nun?, 1932*

Railway stations are magical places, full of promise. They introduce people to a new place at the end of their journey or allow them to make their way to far-away locations, familiarizing them with all sorts of people on the way, from well-to-do business types to those unfortunates forced to live on the fringes of society. There is a nervous energy around them, created by those looking forward to their journey, others trying to get oriented in a place that is new to them, or those who should have arrived earlier and are now faced with the ignominy of running to catch their connection – and likely just missing it.

Many of the processes within the station have assumed a self-evidence that we don't think about: whistles are blown, baggage is handled, tickets are sold, coffee is put into cups, and trains into forward gear. All of these processes have taken time to develop – as did the stations themselves. When railways became a reality in the mid-1800s, and particularly passenger travel promised to be a hugely profitable enterprise, no one really knew how to get people onto trains. Previous forms of transport had typically been small-scale: coaches rarely carried more than a dozen people and time schedules were often flexible. Railway

---

1 A member of the Schutzpolizei, the uniformed branch of the police force.
2 A good example of the sarcastic *Berliner Schnauze*. In the Schutzpolizei, a *Hauptwachtmeister* was a senior police officer with at least 12 years of service who was unlikely to get involved in mundane traffic duty.

coaches, on the other hand, could be coupled to contain several hundred passengers and, in a time when single-track railroads were the norm, trains needed to be at a junction station in time to avoid messing up their timetable.

All of this forced engineers and architects to come up with an entirely new sort of building, suitable to move lots of travellers onto and off of very lengthy objects. Initially, this proved to be a difficult task and the first railway stations were ill equipped to handle people, or trains for that matter. People who arrived did not really know what to do after they collided into those set to depart, and everyone was subjected to the elements.

Fig. 1.2: Hans Baluschek, *Großstadtbahnhof*, oil on canvas, 1904.

Passengers also entered a building unlike anything that had existed before, a space in between two very different worlds. This was particularly true for the terminus, the type of station where the railway lines ended or began. On the one hand, it was a gateway to the city for those arriving; for departing passengers, on the other hand, it became the first stage of a journey. In the words of French philosopher Michel Foucault, the railway station was a heterotopia, a "different", transformative space that contrasted with the world outside, and with its own rules of access and conduct. In the words of historians Jeffrey Richards and John MacKenzie, the railway station:

was [at its most basic level] the nineteenth century's distinctive contribution to architectural forms. It combined within itself in eloquent reflection of the age which produced it both a daring innovative modernity and a heroic and comforting traditionalism.

The modernity was necessary because of the unique demands of this novel form of transport. It involved innovative ways of solving various issues varying from technology to crowd control. The traditionalism, conversely, was a way to reassure people about the reliability of railway transportation and translated into the use of forms or architecture that were sensed to be dependable; closely mimicking and sometimes almost parodying the style of official and ecclesiastical buildings while using historicizing elements. For the same reason, they were often over-engineered and over-sized, and to some extent they still are. Early station architecture also tended to hide the building's essential function and, as time went on and railway travel gained normality and even prestige, this came to be seen in a different light. Firstly, there was no longer a reason to to hide from view what had become a source of pride, but architects also wished for the design and shape of buildings to reflect their function. This became increasingly important in an industrially accomplished nation such as Germany, whose capital, Berlin, is the subject of this book.

Fig. 1.3: View of the Potsdamer Platz from the gallery of the Potsdamer Bahnhof, 1920s.

The Beginning

When the first German railways were built in the mid-1830s, Germany wasn't a country yet. The Holy Roman Empire had been dissolved at Napoleon's insistence in 1806, and its successor, the German Confederacy, only numbered 39 instead of the previous 300+ states. Of those, Prussia was without question the upstart, rivalling the venerable Austrian Empire for influence and military might, so when the first serious attempt at German unification was undertaken during the German Revolution of 1848, the Prussian king Friedrich Wilhelm IV became the logical rallying point. However, after much dithering the king refused, essentially causing the collapse of the revolution. German Unification didn't come about until 1871, but then as a far less democratic project spearheaded by Prussia's chancellor Otto von Bismarck. After three wars against Denmark (1864), Austria (and its German allies, 1866) and France (1870–1871), a new German Empire was founded. With Austria-Hungary absent, it meant that Prussia became entirely dominant: it contained over half of the empire's population, three quarters of its territory and of course its capital, Berlin.

That status, and the wars that brought it about, shaped the way people moved about in the city. Berlin had seen the construction of its first railway station in 1838. The Potsdamer Bahnhof, which formed the Berlin bookend of the railway line to the residence of the Prussian kings, might have been a somewhat inauspicious beginning, but was soon followed by many others. All these were undertaken by private companies; the state played a limited role in the early development of the railway system. Each line initially received its own dedicated terminus, positioned just outside the old city walls, except for the Frankfurter Bahnhof in the east. This was not merely a planological issue; until the 1860s, Berlin still possessed a city wall, which primarily functioned as a tariff barrier (and to keep soldiers from deserting). Property ownership, and hence tax payment, was restricted to citizens and forbidden to outsiders – including corporate entities. Since the Berlin-Frankfurt railway company was the only one to have local citizens on its board, their station could be built within the walls. The other side of the coin was that building a station just outside the walls also meant that the company did not need to pay city taxes.

The decade following the opening of the Potsdamer saw the construction of no fewer than four further stations, serving trains to Anhalt (1841), Stettin and Frankfurt an der Oder (both in 1842), and Hamburg (1847). The Görlitzer (1866–7), the "old" Ostbahnhof (commonly referred to as "Küstriner" Bahnhof, 1867), and the Lehrter (1871) followed some time later. In addition, a whole slew of ad-hoc stations was built and torn down over time.

Fig. 1.4: The first Berlin station, the Potsdamer Bahnhof, ca. 1850. Opened in 1838, the station had already been extended by this time.

Some of these led only short lives: the Dresdener Bahnhof opened in 1875 but closed again in 1882, while the Küstriner functioned for barely 15 years before closing in the same year. In both cases, the reason was the nationalization of the Prussian railways and its subsequent rationalization. The first Potsdamer, Anhalter, and Stettiner stations soon needed replacement by much larger buildings and facilities to accommodate the increase in passengers and goods. Various solutions were devised to connect these to one another and alleviate the pressure on them, including a circular railway, the Ringbahn (1871), and the Stadtbahn, a twelve-kilometer stretch of elevated railway and stations that opened in 1882.

Most termini therefore went through a similar life cycle, influenced by political and social changes: a first, somewhat improvised station in the 1830s and 1840s which was gradually extended; a re-build or extensive renovation in the 1860s or 1870s; some adaptations around the turn of the twentieth century; destruction in World War Two; and demolition during the 1950s and early 1960s.

This cycle followed the major events of German and European history. The first factor was the rise of Prussia as a political and industrial nation, gradually expanding as a result of the wars that led to the formation of the German Empire in 1871. The Görlitzer Bahnhof was a direct result of the Austro-Prussian war, when railway construction was hurried in order to transport troops to the front in 1866. One result of that war was the annexation of the Kingdom of Hannover,

Table 1.1: Dates of the completion, renovation(s), and demolition for the Berlin termini.

Station	Opened	Renovation/ Replacement	Closure	Demolition	Remarks
Potsdamer Bahnhof	1838	1872	1945	1958	Currently Potsdamer Platz Station
Anhalter Bahnhof	1840	1880	1952	1959	Entrance partly preserved
Stettiner Bahnhof	1842	1876, 1903	1952	1962	Currently Nordbahnhof; local station preserved
Frankfurter (Schlesischer) Bahnhof	1842	1870, 1882, 1950, 1987, 1998	–	–	Several name changes, currently Ostbahnhof
Hamburger Bahnhof	1847	–	1882	–	Currently a modern art museum
Görlitzer Bahnhof	1866	–	1951	1975	Currently a park
Küstriner Bahnhof	1867	–	1882	1952?	Entirely vanished
Lehrter Bahnhof	1871	–	1951	1958	Currently Hauptbahnhof
Dresdener Bahnhof	1875	–	1882	< 1911	Site occupied by the Anhalter Postal Station

which in turn enabled the construction of a railway line to Lehrte, and of yet another terminus in the capital, the Lehrter Bahnhof, in 1871.

What Berlin missed out on was the third wave of German station-building just after the turn of the twentieth century. As the country was experiencing its imperial heyday, it saw the construction of massive structures in Hamburg (1906) and Leipzig (1913), the latter still Europe's largest station in terms of surface area. Despite general awareness that the system had reached its limits, no major new stations were built. While a reorganization of railways in the capital was discussed since the mid-1860s and had been partially achieved through the Stadtbahn, unifying all the lines into a single system proved to be a task too massive even for the German Empire. Therefore, Berlin's surface railway infrastructure, in all its imperfection, was more or less complete by the late 19th century.

This book is about the history of these stations and the people that designed them, used them and lived around them. I have chosen to treat them in the chronological order of their opening, with one important exception: the Stadtbahn, the central elevated four-track railway that is still the spine of Berlin's railway infrastructure. Although it opened in 1882, later than any of the main termini, I have decided to discuss it directly after the Schlesischer Bahnhof, since both their history, and that of the entire second generation of Berlin's stations, is so intertwined. The emphasis will remain on the years of the founding of Berlin's eight major termini (plus multiple minor ones) and on their history during the period in which Germany was officially known as the German Empire (Deutsches Reich): the Imperial, Weimar, and Nazi years. Berlin's post-war history had its own, unique dynamic, one in which Berlin's old termini played a negligible role. Allow me to sketch some perspectives on the subject of Berlin's railway development before discussing these termini in detail.

Assumptions

Berlin's early railways stations suffered from the consequences of two miscalculations, both rooted in Berlin's explosive growth during the 19th century. When the Potsdamer Bahnhof opened in 1838, the city boasted just shy of 300,000 citizens – a big city by early nineteenth century standards, but nothing compared to what it was to become over the next century. By the time of German unification in 1871, that number had already swelled to over 800,000 people, but it really took off after that, reaching one million only five years later, and two million around 1900, the quickest growth of any city at the time after Chicago. The numbers become even more impressive once we factor in the similar growth of Berlin's satellite cities, such as Charlottenburg and Schöneberg. When these formerly independent entities were amalgamated into Greater Berlin (Groß-Berlin) in 1920, the city's total population nearly doubled. During all that time, city officials struggled to keep pace with the requirements of an ever-growing populace, and the railway companies were as overwhelmed as everyone else.

The first false assumption the pioneering companies made was that the largest profits were to be earned in passenger travel. When after a few years it turned out that all the various goods needed to feed a relatively isolated metropolis was going to be the bigger earner, the city already possessed a sizeable railway infrastructure with wholly insufficient installations for coping with all that freight. This meant that it often had to be handled at passenger stations, crammed in between existing construction, or moved to the edges of the city, from where further transport might clog up street traffic.

The second big mistake was the expectation that long-range travel would make up the majority of passenger movements. Rather, however quickly demand

Fig. 1.5: Waiting crowd at Stettiner Bahnhof, 1935. Photo by Willy Pragher.

for more distant travel grew, it was soon overtaken entirely by the need for local and regional means of transport as the city sprawled and workers needed to be transported. When the railway authorities began to address this structural problem, it was with hesitancy. Snobbery also played a role in the strictly regimented Prussian society; long-distance travel involved sophisticated people visiting faraway lands, while local railways transported dreary commuters and grimy laborers to their places of work.

While Berlin quickly became a very big city it was also a relatively isolated one: unlike in the overcrowded Rhineland or the Main area, the nearest major cities were quite far away, which made travel to them arduous, often expensive and therefore something to be avoided if possible. Consequently, and like most metropolises, Berlin became more or less its own nation, limiting the flow of people away from it while simultaneously increasing the flow of goods towards it. Those early developments continued to haunt the city for a long time.

Bahnhof to Bahnhof

An added problem was that those for whom Berlin was not their final destination were doomed to a considerable amount of cross-city travel. Early on, Berliners considered ways to connect their stations, but the solutions they invented for inter-station travel initially owed more to improvization than to planning. The first such attempt was the Berliner Verbindungsbahn (Berlin connector train), which opened in 1851. Rarely did a train service have "afterthought" written over it to the same extent as this single-track, street-level steam train, which ran around Berlin's center, adding further noise and smoke to an already over-crowded and polluted city; it was not even primarily meant for passengers (apart from occasional troop transports), but rather to carry goods and supplies between stations and to industry. Regularly, whole avenues were blocked by the long trains of the *Verbinder*, and complaints about pollution were rife. In 1864, it was restricted to night hours, which meant that while the streets were at least naviga-ble again, everyone's sleep was ruined by the noise.

Still, Berliners should be grateful to the Verbindungsbahn because its short-comings prompted the construction of the Ringbahn, a mostly elevated railway circling the city at some distance and connecting the various termini through the outer city. Its opening in 1871 meant a welcome goodbye to nightly steam trains in the city streets (except for a short stretch still used for coal transports). A few years later, the Ringbahn was extended to a full circle, although all local passen-ger trains had to go in and out of Potsdamer Ringbahnhof from the south. Today, we know it as the S-Bahn ring.

Ten years after the Ringbahn, the Stadtbahn was added, an elevated railway crossing the city center. This is the railway passengers are likely to use when visit-ing the city today, and it serves Zoo, Hauptbahnhof, Friedrichstraße and Alexander-platz stations, among others. It opened in 1882, connected the Frankfurter Bahnhof (which was turned from a terminus into a through station, and renamed the Schle-sischer Bahnhof) to Charlottenburg, and went straight through the heart of the city before meeting up with the Ringbahn. However, it connected to none of the older termini with the exception of the Lehrter Bahnhof. Here, a Stadtbahn station was built on top of the existing lines entering the station.

It is here that Berlin Hauptbahnhof now stands, on the site of the old Lehrter Stadtbahnhof and Lehrter Bahnhof. Anyone arriving or departing from the top platforms of the new station will have noticed how cramped it still is, since it con-tinues to suffer from the restrictions of a nineteenth-century system – and an ex-tension of that capacity is all but impossible in such a densely packed city. In addition, it now also uses a north-south tunnel. It is still clumsy, but a lot better than having to fight your way through a city to get from one station to the next.

Fig. 1.6: Entrance to the Lehrter Stadtbahnhof. Hermann Rückwardt, 1882.

That fight was made a lot easier over the years, of course. Berlin's first horse-drawn omnibus service was introduced in the 1840s and ran between the Pots-damer Bahnhof and Alexanderplatz. A horse-drawn tram service followed in 1865, to be replaced by an electrical one after 1895. The opening of the first Hoch-bahn (elevated railway) in 1902 was, however, a more substantial innovation. Un-like the trams and buses, this mode of transport was entirely separated from street traffic and could therefore be both faster and more reliable. Over the next decades the system of electric overground and underground trains was quickly expanded into what we know today as Berlin's U-Bahn system, and greatly facili-tated crossing the city.

Under Pressure

It was something of an accident that the capital's railways developed the way they did, the consequence of being built on private enterprise combined with Berlin's explosive growth. To make matters worse, the number of travelers tended to in-crease at a far greater rate than the general population. In 1840, the total number of long-distance passengers was around 300,000; by the end of the century, it had increased to over 4 million. But the real explosion came after 1900: to 10 million in 1913, and on to 40 million in 1922.[3] However, that number still accounted for a mere one fiftieth of German long-distance rail movements (at 2 billion). It was also only a fraction (one thirtieth) of local travel, including suburban rail but also subways, buses, and trams.

Fig. 1.7: Carriage with holiday luggage arriving at the Stettiner Bahnhof, 1904.

Thanks to the increased accessibility of the railways, travel had also become increasingly seasonal; huge spikes occurred around Easter and particularly Pentecost. Christmas was not as significant a holiday at that time as it is today; moreover, it took place in the dead of winter, when people preferred to stay at home. The biggest factor, however, was tourism, which initially consisted mainly of better-off families visiting the Baltic shore to cool off during the sweltering Berlin summer. As time progressed, improved infrastructure made day trips easier and more popular, which ramped up the number of journeys even further. The Anhalter and

---

3 Although it needs to be said that the 1921–1922 numbers were inflated; Germany's hyperinflation of the early 1920s also meant that travel had become very cheap because ticket prices did not keep up with monetary devaluation. When the Gold Standard put a stop to hyperinflation in 1924, passenger numbers flattened out considerably.

Stettiner served as the main stations for holidaymakers; the latter consequently became the busiest station in the city.

All of these movements put an enormous strain on Berlin's transport infrastructure. Lines and stations could not be built quickly enough, as projections of traveler numbers continually proved to be far too modest. Some have suggested that Berlin's railway system was planned foremost with military use in mind but, although this might have been intended, the truth is that their development owed more to happenstance than it did to coherent planning or the agenda of any particular interest group. While military considerations were always important in Prussia, no plan could be devised that compensated for the massive increase of travel movements; Berlin's transport history is therefore one of continuous improvisation, adaptation, and sometimes outright panic. The fact that the military decided to construct its own railway system, as the public one was so over-used, is telling (see chapter ten).

The need for better urban (and national) planning was one of the reasons why the Prussian state started to consider nationalizing the various railway companies. Already in the 1860s, efforts were undertaken to bring them under state control but, after wholesale nationalization failed for both financial and legal reasons, the companies were nationalized one by one, an arduous and expensive process that could not be completed until 1884. From that year on they were all part of the Prussian national railways, which made it possible to take a broader look at Berlin's railway landscape and plan for better integration of services. After the First World War, the Prussian Railways were integrated into the Deutsche Reichsbahn, which covered all of Germany and made it possible to consider Berlin's railways as part of a national infrastructure. Over time, this gradual organizational defragmentation of the system led to a number of plans to re-design the capital, which overlapped and inspired each other to a surprising degree (see chapter eleven).

Prussia's army, which played a dominant political role, had a hand in most decisions about transport infrastructure in Berlin, including nationalization. Initially sceptical of the railways, their enthusiasm quickly grew as the strategic advantages of rapid transport of personnel and material became apparent. However, the chaotic nature of Berlin's railways caused the armed forces to install their own railway infrastructure (see chapter ten). The military deemed them to be too important to leave in the hands of private enterprises, although many of these companies' problems were also partly a result of that same army's interference. For example, new rail lines sometimes needed to be built across expansive parade grounds on the edges of the city, and the compensation arrangements could be financially crippling. Nonetheless, military considerations were an important factor in the nationalization of Prussia's railways in the Royal Railway Directorate (Königliche Eisenbahndirektion).

As more and more lands were added to the kingdom and the empire, its railway network grew, as did the importance of Berlin as the center of an extensive web of rail links. This contributed to the rapid growth of the city itself, which put a great deal of strain on its transport infrastructure, both for passengers and goods. Contrary to, for instance, London or Paris, Berlin's termini tended to be built under-sized, a consequence of the companies' lack of financial resources but also of a typical Prussian ideal of austerity in public expenditure; a terminal needed to be grand, but it was never to seem wasteful. This helps to explain why the buildings, particularly the train sheds, were not nearly as imposing as the large terminal stations of Paris or London; the largest of the Berlin stations in terms of surface area, the Schlesischer Bahnhof (today's Ostbahnhof), would not even have featured in either of the other two capitals' top-five.

Nationalization of the railways generally improved their financial position, but the spirit of penny-pinching long persisted. Also, the process proved lengthy and expensive for the state; by the time it had been completed, the expense, and a new financial crisis, prevented further grandiose improvements of the system. Despite capacity problems at several stations, others were taken out of service despite being perfectly serviceable and sometimes newly renovated. In 1882, this happened to no fewer than three stations: the Hamburger, Küstriner and Dresdener stations were all closed.

Evolution

The difference between the first and last specimens of Berlin's first generation of stations was already substantial. Whereas the Potsdamer was little more than a platform with a single track next to it, the beginnings of a more modern, integrated transport terminal can already be recognized in the Hamburger, built about a decade later. However, for a long time Berlin remained a city designed to arrive (and stay) in or depart from. It took until the 1880s for the first major through station in Berlin to be opened, and even then, it was an adapted terminus. Part of the reason for this lay in the logical status of the national (and later imperial) capital as the endpoint of a railway line.

The railway terminal therefore became the standard, also because it offered all sorts of advantages for an independent company. The station was a huge advertisement sign for the company in the city it served. It could control everything for its passengers: not only trains but also facilities that would serve them before and after their journey. For the city, the existence of so many stations was a problem, of course, since customers had to be very aware of which destination was served from which station. The need for more cohesive urban and national transport planning was one of the reasons the Prussian state decided to nationalize these railway companies.

Termini are usually impressive buildings and particularly satisfying from an architectural perspective since they offer such an identifiable "face." They have often been referred to as "railway cathedrals," or "cathedrals of steam"; what unites them is a desire to impress the churchgoer and passenger, respectively. It is not entirely coincidental that many a railway architect also distinguished themselves in the construction of churches. For example, the designer of the mighty Anhalter Bahnhof, Franz Schwechten, was also responsible for the Gedächtniskirche, the church at the beginning of of Kurfürstendamm.

Fig. 1.8: A true railway cathedral: Anhalter Bahnhof's train shed looking towards the reception hall, 1880.

As the dimensions of the rail industry grew and procedures became standardized, so did stations; however, this did not always turn out to be convenient. The history of Berlin's stations is therefore also a history of adaptation of their concept to new realities.

The most evident example is the strict separation of arriving and departing passengers that became the norm with the construction of the Frankfurter Bahnhof in 1842. That station had two buildings on either side of the tracks, one serving arriving passengers, the other departing ones. This involved lots of walking, a cumbersome infrastructure for baggage, and more staff; moreover, it added logistical restrictions to an already complex operation. However, because this setup had been laid down in law, little could be done until the Prussian Railways were allowed to change their regulations in 1893 and install platform barriers. This was also done to control admission; until that point everyone was able to wander about in the station, which caused the platforms to be even more crowded and encouraged crime; with barriers installed, operations could be simplified, since everyone needed a ticket to enter the platforms. This could be either a travel ticket or a separate platform ticket (*Bahnsteigkarte*), an invention that only allowed people to be on the platform, for instance to welcome an arriving friend or relative.

Another influential factor in the development of railway stations was the progress of technology. The first stations, for example, were set up with a turntable, to turn rolling stock around and allow them to leave using a different track to the one they had arrived on. These soon proved to be impractical, as they limited the size of rolling stock and required a lot of space in front of or inside the building. Often, they were replaced rather quickly; initially by a transfer platform, on which the locomotives could be pushed to a different track, and later by switches inside the train shed.

Finally, we come to an element that is not always as evident: convention. The reasons why the first stations were built in an architectural style reminiscent of contemporary official buildings and on historical revivalist templates have already been discussed. However, there were other conventions that also determined architectural decisions. One is the *Berliner Breite* ("Berlin Width"), a standard of 37,66 meters for the façade of official buildings, which continued to constrain the dimensions of Berlin's stations until it was finally abandoned with the construction of the Anhalter Bahnhof in the late 1870s. By that time, architects had done away with other restrictions as well, and station architecture had become an independent genre, with an individual stylistic language all its own.

A Berlin Station, c. 1900

Let us sketch a typical station experience for a passenger setting off for a destination outside the capital around 1900. Depending on their means and fitness, they might walk, take a horse-drawn tram or a Droschke (horse-drawn taxi) to the station, which they would likely enter on the right side, where the departure platform was located. Some stations still possessed separate doors for departing and arriving passengers, but others allowed them to enter through the front; here, they might deposit their luggage at a designated desk for a fee of half a Mark per 10 kilograms. Other facilities could usually be found here as well, such as a mail office and financial services; in some stations, even medical facilities were available. In the station hall, passengers could also purchase a ticket for their journey at a counter. However, usually the largest part of the building still consisted of offices and living quarters for railway personnel.

The differences in ticket prices were notable. Prices were set by distance; in 1899, a ticket for the 57-kilometer journey from Berlin's Friedrichstraße station to Brandenburg cost 5,13 Marks in first class, 3,40 in second, 2,28 in third, and 1,14 in fourth.[4] On the eastern railways, the fourth class would consistently be the most-used one; on the railways towards the west, however, a third-class ticket was the most common. First-class carriages were sumptuous, and care was taken that a decrease in class neatly corresponded with a decrease in comfort. In one case, the railway company decided to leave out glass windows in their new fourth-class carriages to make the experience clearly a worse one from that in third class.

Stations usually contained waiting rooms for each class of passenger, plus at least one separate room for ladies. Sometimes the first and second class rooms were combined, and sometimes also the third and fourth. Care was taken to place the first class rooms at the most advantageous position in the building, furthest away from the noisy areas such as the baggage counter, and closest to the one waiting room that you could not buy your way into, reserved for the *Allerhöchste Herren*, the "very highest gentlemen", i.e. the Royals and their court. Typically for Prussian society, Berlin's stations were strictly socially regimented. Military personnel, who purchased their own, distinct tickets, were therefore obviously assigned a travel class according to rank.

---

[4] Although the conversion of historical currency is tricky, one way of calculating is to set it against the price of gold. In that case, these figures should be multiplied by 20 to get the current amount (2024) in Euros. If we use labor cost, however, the multiplication factor comes to around 75 (calculated from https://www.historicalstatistics.org/Currencyconverter.html).

Fig. 1.9: Great hall of the Stettiner Bahnhof, (left) baggage counter next to the stairwell to the train shed and (right) entrance and ticket counters, 1903.

Needless to say, the facilities for those traveling in the two highest classes were of a rather different standard than those in the bottom two. Some first-class restaurants, such as the one in the majestic Anhalter Bahnhof, gained a reputation for their excellent kitchen and service; by contrast, third-class waiting rooms could be a fairly dingy affairs, often with a dubious reputation. For instance, the third-class room in the Schlesischer Bahnhof (today's Ostbahnhof) was well-known for being a place where the infamous *Ringvereine* conducted their crime business (see chapter five).

Initially, people could only enter the platforms through the entrance and arrival buildings. When platform barriers were introduced during the 1890s, travelers needed to wait at the head of the platform until they were opened. This also made it possible to open the terminal platform to commercial services, such as store kiosks and snack stands, and to centralize information facilities.

Having waited for their journey in their designated waiting room, passengers then boarded their train to travel in one of the four classes. Initially, there were only three, but beginning in 1857 the Silesian railway (to Breslau) and Ostbahn (Eastern Railways to Königsberg) introduced a fourth class as a budget option. That this was done by the Ostbahn is significant because it connected Berlin to the poorest part of the country; the Anhalter Railway was the last one to introduce a fourth class ten years later. However, fourth class carriages were never added to express trains as the surcharge for these rendered them too expensive anyhow.

The share of first-class passengers always remained minute. In 1900, the Prussian State Railways transported about 2 million of them out of a total of around 580 million passengers. With 57 million, second-class passengers were slightly more numerous, but their numbers still remained insignificant in comparison to those traveling in third (279 million) or fourth class (243 million). Consequently, first- and second-class carriages often drove around empty while third and particularly fourth-class ones were filled beyond their capacity – very visible demonstrations of privilege and poverty that did not go unnoticed at the time.

Planning

Neither did the way in which the system remained under pressure. It had long been recognized that the setup of the capital's railways was quite inefficient, with ideas about revision going back as far as 1865. Through the years, several plans were formulated for a large-scale reorganization of transport in the city. Arguably the most famous of these is Adolf Hitler and Albert Speer's construction of a new capital, the *Welthauptstadt Germania*, on top of Berlin. Speer integrated his megalomaniacal building plans with existing ideas for two massive stations at the north and south end of the new city center. Apart from minor construction and major destruction – leveling entire neighborhoods of the city – the war and then the well-deserved end of the national socialist *junta* scuppered any chance of the project being brought to completion.

In effect, the onslaught of proposals and speculations worked to effectively stall decision-making (see Chapter eleven), and even when a consensus was on the horizon, two world wars worked against its execution. In a way, the Second World War and its consequences solved most of the capacity problems, while the Cold War, in which Berlin "enjoyed" a rather exceptional place, sealed the fate of most of Berlin's termini. The war brought wholesale destruction to the German capital, of which its railway installations bore much of the brunt. Some stations could be used again in the summer of 1945, but others never saw traffic again.

After the war, there was little enthusiasm for rebuilding Berlin's railway infrastructure in its old form. To begin with, The reduction of Germany's territory after the war entailed the removal of many previous destinations, and the escalating division of Berlin meant that fewer and fewer passengers could use many of the old stations. For that reason, buildings such as the Anhalter, Stettiner and Görlitzer Bahnhof no longer served a purpose.

Fig. 1.10: Disused destination signs at the ruined Lehrter Bahnhof, 1957.
Photo by Manfred Niermann.

Much of what could have been rebuilt was demolished, nonetheless. While architecturally lamentable, the disappearance of Berlin's termini is also understandable for a number of reasons. First, after the partition of the city between the four Allied powers most stations lay in different zones from the tracks that fed them, which became particularly problematic after 1951 when citizens of the Soviet-backed German Democratic Republic were formally (and inefficiently) forbidden from entering West Berlin. The Anhalter, Stettiner, and other stations were still in use even in their bombed state, but people traveling to, say, Magdeburg could no longer get to the place where they were supposed to begin their journey.

A second convincing reason has to do with the intrinsic shortcomings of termini. For the traveler, the setup of all tracks stopping in front of a big hall can have a host of advantages; it has an intuitive obviousness, all the facilities can potentially be grouped together for the benefit of everyone, and wayfaring and information supply is generally quite simple to organize efficiently. There are drawbacks as well, though; traversing a terminus generally involves a lot of walking, because passengers always start on the platform near the tail end of their train. The first termini also did not exploit potential advantages of their setup very efficiently, as arriving and departing passengers were strictly separated. Changing a train could therefore mean walking around the building.

Besides, regardless what advantages termini might have offered to passengers, they were a nightmare when it came to transport logistics. To begin with, platforms and tracks couldn't be used nearly as efficiently as in through-stations; usually one half of the station was reserved for trains arriving and the other half for those departing, effectively halving the building's capacity. Driver cabins (and in the past, engines) needed to be on both ends of the train since driving in and out had to be done from different ends. Parking trains after service was also much more complicated because it interfered with regular traffic. All those operations also required a lot of space, and the presence of the enormous rail yards in the city became increasingly problematic as the city became busier and more dense.

As imposing as they might have looked, the capacity of these stations was not that impressive to begin with. As we saw, most of the Berlin stations were not particularly large; it was just that there were a lot of them, which made their aggregate capacity considerable. However, at just over 50,000 square meters of train shed surface combined, they still fell 10,000 short of Leipzig's massive main station (1913) alone, and only just topped Frankfurt's and Munich's.

Finally, the post-war atmosphere was determined by the wish to forget about the war and what had led up to it, and a new generation of architects had come to despise historicism as a symbol of the various old and suspect regimes. It was an ideological aversion aided by the career opportunities that the replacement of such structures provided. Hans Scharoun, the architect of Berlin's Kulturforum in the late 1960s and an important voice in the city's heritage institutions at the time, was at least partly responsible for the demolition of the Görlitzer Bahnhof by calling it "architecturally worthless". His voice was hardly a unique one at that time. Additionally, the rising dominance of the car, and car-centric city planning after the war threw the future of rail travel as a whole into doubt. As West Berlin saw an inordinate amount of highway construction, its railways crumbled.

The Remains

What remains today is not much. The Potsdamer and Küstriner Bahnhof have entirely disappeared, and in the case of the Stettiner and Görlitzer only some auxiliary structures remain. The Schlesischer Bahnhof is now Berlin Ostbahnhof, having gone through two rebuilds and three re-namings since the war. The Hamburger Bahnhof, already out of service well before 1900, then turned into a transport museum and finally into the art museum it lives as today, has survived its tribulations comparatively well. The Stettiner has disappeared except for the Vorortbahnhof (suburb station) that had ironically already become redundant before the bombs fell and survives as an entertainment venue. Their disappearance has fundamentally changed Berlin.

Fig. 1.11: From Gateway to car park: the Anhalter Bahnhof's remains, 2024.

But how? What did these structures mean to the city? Why and how did people travel from and to them, what was their influence on the surrounding area and the city as a whole? And why does this part of Berlin's past still conjure up such nostalgia? There are exceptions, of course. Consider Grunewald Station – not a terminus strictly speaking, but very much representative for the murder of Berlin's Jews, most of whom started their journey to the death camps here during the Second World War (see chapter ten).

Perhaps none of Berlin's old stations symbolizes this nostalgia more than the Anhalter Bahnhof, the portico of which survives near the eponymous S-Bahn station. Once Berlin's most prestigious station, its flawlessly restored, remaining ruin serves as a reminder of both the horrors of the Second World War and the developments afterward. There are plans to reconstruct it as an entertainment venue or as the entrance to a new Berlin Exile Museum, plans that still have yet to get off the ground. The mere fact that the fate of the lost termini of Berlin is still being discussed so long after most of their demise tells us something about the places these sites still hold in the public imagination.

2 Potsdamer Bahnhof: One for the Price of Three

Fig. 2.1: Potsdamer Bahnhof on a colorized postcard, before 1922.

There is a place in our turbulent Berlin that combines securely enclosed narrow spaces with a view of endless expanses in an unusually sympathetic way. This is the first-class waiting room in the Potsdamer train station.

A small room with a few whitish tablecloths and worn, red upholstered furniture from the time of the duck plague. Only a few visitors, the iron core of every waiting room: the lady with the suitcase, who always looks at the time and never leaves. Then the syphilitic elderly gentleman, who occasionally goes to the buffet; the maid automatically reaches for the cigar box, but he just wants a stamp. So much for the inventory.

But outside in the hall, the locomotives snort and pant, the wheels screech, the sirens toot the chorale of the nameless expansions. Through the frosted panes you see hasty shadows scurrying. Screams, farewell calls, endless roaring.

Close your eyes, and between the pounding and the noise the mental landscapes smooth out (that's how one says, more correctly: one writes it), the moving mountain ranges, the peaks and chasms of your thoughts sink, your emotional life flattens out genially into an immeasurable surface. It's good, it's wonderful. This is more than a dream, with its disturbances that play into one's waking hours. This is that all-flattening stupidity that is most likely identical to complete bliss. – *Lucius Schierling (pseudonym of Carl von Ossietzky), 1924*

Diplomatic hotspot, cultural hub, war-torn battleground, Cold War wasteland, and revived urban center, the Potsdamer Bahnhof in Berlin saw it all. And al-

https://doi.org/10.1515/9783111381879-002

though it generally is not perceived as Berlin's most important pre-war station, in a way it was.

First of all, it was the oldest one, the bookend of the first Prussian rail line between Potsdam and Berlin. Secondly, it was in the best place: within walking distance of the Brandenburg gate and Reichstag, and almost right up to the Potsdamer Platz, the city's busiest intersection. The Leipziger Strasse, leading away from the square and the station, ran straight into the city center, crossing the Wilhelmstraße – the home of government and diplomacy – and the Friedrichstraße, the city's prime shopping avenue. For many Berliners, Potsdamer Bahnhof was the most important station in the city in a practical sense; it was where traffic from Potsdam and Magdeburg started and ended and a major commuting hub.

Drafty Days

The building of the Potsdamer Station opened on 29 October 1838, when the first rail line between Potsdam and Berlin was inaugurated.[5] Built and financed by a private enterprise that was created for the occasion, it was something of a makeshift affair, partially because it had been constructed in some haste, but mostly because no one was really sure yet what a station building was supposed to look like.

Like most early stations on the European continent, it drew on English examples, particularly the first "real" passenger railway station, Liverpool's Crown Street. Opened in 1830, the terminus of the Liverpool and Manchester Railway consisted of a modest, two-story building next to a covered set of tracks. However, it contained all the essential elements of later and modern stations: a building to receive passengers and offer them services, such as ticket sales and baggage handling, a platform next to the tracks and a train shed to cover both the platform and tracks. What it did not possess was an easily identifiable style of architecture; the reception building looked like any contemporary building in the city.

Similarly, the design of the first Potsdamer Bahnhof drew on the style of official buildings in Berlin, such as the new customs office designed in 1832 by Friedrich Schinkel, Prussia's most prominent architect at the time. Thus, Schinkel's neoclassical style became part of most early Berlin railway stations.

We cannot be entirely certain what the station looked like on its opening day; the various depictions of the station sometimes contradict one another and are often

---

5 That first line pretty much followed the route of Berlin's present-day S1 line up to Zehlendorf, only to continue straight on to Griebnitzsee station and, finally, Potsdam.

Fig. 2.2: The Potsdamer Bahnhof shortly after opening.

idealized, as such drawings tended to be at the time. Like Crown Street, it was of a "building-next-to-a-line" setup; of the three tracks, the one next to the building was entirely dedicated to passenger transport. The complex was more generously proportioned than its English example, however. Firstly, the authorities forced the railway company to purchase a large plot because they wanted to keep some distance between the station with its unsanitary activities and the surrounding buildings. The building itself, moreover, was going to be used by the royals to reach their palaces in Potsdam, and therefore needed to be built to a representative standard and size.

Whatever it looked like, it did not do so for long; almost from the beginning the station was subjected to various, seemingly random adaptations and extensions, and the rather minimal first reception building was extended by separate structures over the 1840s to house proper waiting rooms, ticket and baggage offices, facilities for rolling stock, and a goods station.

Capacity was not the only problem; comfort – or rather, lack thereof – was at least as big a complaint levelled against Prussia's first terminus. This first batch of Berlin termini had only their platforms covered. Crown Street Station's tracks had been covered, but that made it difficult for the smoke and soot from the engines to escape, causing inconvenience and even danger to passengers. The obvious downside of leaving out such a roof was that travelers were thus exposed to

Fig. 2.3: The Potsdamer Bahnhof in the 1850s, half of a stereo photograph. View from the Potsdamer Platz.

the elements, including (in the case of unfavorable winds) a heavy dose of engine smoke. It was not until the 1850s that new construction methods allowed for the building of much higher and broader halls, built using cast iron and later steel beams. Engine smoke could rise to the top of the hall, far above the passengers who were now also properly insulated from bad weather. The first of these new stations were built in London and included King's Cross (1852), Fenchurch Street and Paddington (both 1854), and, most impressive of all, St. Pancras (1868).

Before that, various solutions were attempted to protect the passengers; the Hamburger Bahnhof (opened 1847), for instance, allowed engines to exit the station hall at the front, where a turntable could be used to turn them around and return, using the other track. While this protected travelers from most of the elements, it still made the hall an uncomfortably drafty place to be. For arriving passengers, things were even worse, at least initially; there were no facilities to get them comfortably off the train. Instead, they were deposited on the platform and had to find their own way into the city.

For decades, the station area became a more or less permanent building site, as the railway company grappled with the growth of its business. The first real improvement came in 1846, with the installation of a canopy over the passenger platform, which gave the travelers at least some protection from the elements. About a decade later, an observer already noticed that only a few of the original buildings were still left standing. The original reception building remained, although it was continually extended; in 1862, a new baggage facility brought some relief, but with the constant increase in passenger numbers, the situation remained entirely unsatisfactory.

New Methods, New Style

Fig. 2.4: Potsdamer Bahnhof under construction, late 1871.

By 1861 the authorities had their fill of the ramshackle complex at Potsdamer Platz that now awkwardly handled over a million people a year, far more than it had ever been designed for, and they demanded a full re-development of the station. After some back-and-forth the railway company acquiesced, and set a host of Baumeister to work, led by the company's director Julius Quassowski.[6]

The design Quassowski and his collaborators came up with used an architectural style known as the *Rundbogenstil,* ("round arches style") which was commonly used on Berlin stations and in other official buildings at the time. As was usual throughout nineteenth century architecture it used elements from historical building designs, in this case predominantly Romanesque (which explained the round arches) and renaissance architecture. It can be seen as part of the German national movement of the early nineteenth century: a style expressly conceptualized as a specifically German style of building, continuing on a neoclassical basis and opposing the prevalence of (English) neogothic designs of the time. Particularly recognizable was the use of "eyebrows," arches above sets of windows or doors. While the use of these round arches was the style's standout feature, other elements also returned, such as columns and galleries. Furthermore, it included specifically "German" elements, particularly features from medieval architecture such as rustications, crenellations, or small towers.

Fig. 2.5: *Rundbogenstil* adopted in Munich's Station, 1854.

---

6 The actual architects were Erwin Theodor Döbner, Karl Weise and Anton Sillich. The term Baumeister was typical for Germany, and combined the capacity of an architect, engineer, and sometimes contractor (although increasingly more the former and less of the latter).

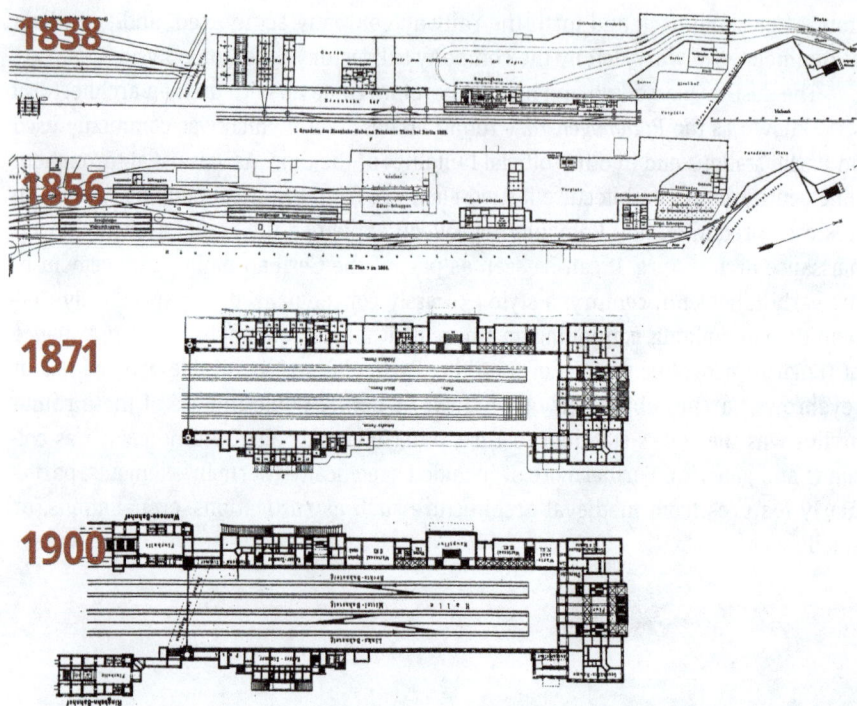

Fig. 2.6: Four phases of the Potsdamer Bahnhof from beginning to end. Top to bottom: the original station (1838), with extensions (1856), the new station (1873), and with the two side stations (1900). The north is to the right. In the upper two images, the old city wall is still visible.

Berlin had been relatively late in the adoption of the Rundbogenstil in its railway architecture. Some of Germany's earliest stations, such as the ones in Leipzig and Munich, had been built incorporating it at a time when stricter neo-classical designs were still common in the Prussian capital. However, that changed during the second wave of Berlin station-building in the 1860s, when Rundbogenstil elements became standard. In fact, the city's fondness of the style would remain until after the Second World War, when the reconstruction of Ostbahnhof in 1950 still incorporated some of its stylistic elements.

Construction began in early 1870 on a new Potsdamer Bahnhof, built on the same principles as the London termini, with a single, big hall using steel beams and a roof covered by glass to let in light. Where the old station had used a single platform to serve all traffic, the new one applied the same strict separation between arrivals and departures that had become common on the other Berlin termini. The whole railway was raised as well, to separate it from street traffic. This

Fig. 2.7: Entrance hall of the departures wing of Potsdamer Bahnhof at the time of opening, 1872. The train shed can be seen through the windows on the right.

also caused some controversy, since it visually interrupted the *Generalszug*, a system of avenues and streets constructed between Charlottenburg and Kreuzberg and named in honor of Prussian generals from the Napoleonic Wars.

The new, much larger and grander building was opened on 30 August 1872 by the arrival from Gastein (by train, of course) of German emperor Wilhelm I, who had recently ascended to this position. As it was being built, the station had gone from being "just" a Prussian rail station to being the main transport hub of the new German capital, and both emperor and Reich chancellor Bismarck had shown great interest in the building process. However, at the time of opening it had not been entirely finished, with the platform the aging Emperor alighted on being the only one completed. The royal waiting rooms were ready, but Wilhelm traipsed through them without looking and made his way – of all things – to a horse-drawn coach that was

Fig. 2.8: View of the Potsdamer Bahnhof and its surroundings, by Julius Jacob and Wilhelm Herwarth, after 1891. The single tower of the Wannseebahnhof is visible to the right of the main station.

to transport him to his city palace. Common travelers had to wait for a few months more until the station was opened to them as well, on November 1.

Once finished, the new building was certainly an improvement. Much bigger than the old station, it also offered far better waiting rooms and facilities and, of course, looked like a proper place to arrive in a city that was increasingly beginning to regard itself as an international metropolis.

However, it also copied some shortcomings of earlier stations, which were already criticized at the time. In Prussia, it had become usual to separate arriving

from departing passengers. The recently opened Küstriner (1868), Görlitzer (1867), and Frankfurter (1869) stations were therefore "double" stations in practice, which made some sense for stations that served a single line, but not for a facility operating several, such as the Potsdamer. The ornate, neo-renaissance front with its two arched galleries was useless for departing passengers, who were required to enter at the right of the building; this section also contained the ticket and baggage offices, and the waiting rooms for four classes of passengers. Arriving passengers could leave the stations through the front or the left side, where a large space was reserved for coaches to carry the travelers to their destination in the city.

Things were different for those of royal birth or in the service of the royal household. As the "palace's station" the Potsdamer contained separate lounges in both the arrival and departure wings for the *Allerhöchste Herrschaften*, initially placed at a proper distance from the regular waiting rooms.

The separation of arriving and departing passengers also meant that services could not be centralized. Having left your luggage at the baggage office upon departure, you had to circle the building to pick it up again, or wait until it had been brought to the office on the arrival side. Initially, it was also not possible to cross from the departure to the arrival platform, or vice versa.

After 1893, a central access and departure point was created at the head of the tracks at most stations, including the Potsdamer Bahnhof. This made checking tickets much easier, but also generally forced passengers to walk longer distances. The front building remained largely unused for passenger services, containing offices and (mostly) living quarters for railway personnel.

From One to Three (or Four, or Five)

In a repetition of the first building's problems, the new station soon proved far too small. Another rebuild was hardly an option; the station was still fairly new, the Potsdam line had become a state enterprise, and, furthermore, the economic depression of the 1880s made further state expenses controversial. So, rather than tackling the problem head-on by initiating yet another wholesale reconstruction, the old habit of improvising smaller extensions was again adopted; by the turn of the century, the Potsdamer Bahnhof had turned into a three- (or even four, depending on how and what you count) station complex.

In the middle, and most eye-catching, was of course the old long-distance station. While not the largest, it was situated in a crucial position. The opening of a new Anhalter Bahnhof in 1880, situated four hundred meters to the south-east, took away some of its luster as an international hub, but certainly not all of it. To the left and right of the station, two suburban stations were created in 1891 to relieve pressure on the main building, In the east there was the "Ringbahnhof," which consisted of two stations: one platform was reserved for the Ringbahn, the suburban circle line to areas further out; the other connected to the Stadtbahn (also via the Ringbahn), the line traversing the city center (see chapter six).[7] To the west of the station,

---

7 The terms "Stadtbahn" and "S-Bahn" are often used interchangeably, which sometimes causes confusion. The Stadtbahn was the elevated stretch of railway, opened in 1882, that crossed Berlin's inner city from Schlesisches Bahnhof (today Ostbahnhof) to Charlottenburg, while the S-

Fig. 2.9: Inside the Wannseebahnhof, 1915.

and near its main side entrance, the so-called Wannseebahnhof, another commuter station, transported people to and from the stations on the Wannseebahn, which partly ran on Germany's oldest line to Potsdam and then through affluent suburbs to Wannsee station, situated next to a large lake in the Havel river.

Socially, the difference between both commuter stations could not have been bigger. The Wannseebahnhof was where the most prominent Berliners commuted to on a daily basis: civil servants, diplomats, and academics, who had found a home away from the noisy and grimy city in the leafy towns of Zehlendorf, Steglitz, and Lichterfelde. Among them were special "banker's trains" (Bankierzüge) that transported financial professionals directly between the wealthy satellite town of Zehlendorf and Berlin's city center.

By contrast, the crowd that used the Ringbahnhof was of a more lower-middle-class and proletarian nature. Many of them did not end their daily journey at the station; they took other services to their place of work or walked there. That walk started out at the station itself since it was situated at the back of the

---

Bahn is a means of transport which sometimes uses the Stadtbahn but is far more extensive. The precise origin and even meaning of the term "S-Bahn" itself are disputed.

Fig. 2.10: The Potsdamer Bahnhof complex in the late 1920s. The Ringbahnhof is to the left, the Wannseebahnhof to the right of the main building, to the left of which is the entrance to the U-Bahn station and the dome of the Haus Vaterland.

main station; having left the Ringbahnhof, commuters needed to walk the length of the main station to their connecting service, a distance of a quarter of a kilometer. A tunnel had also been installed to allow for easier passage to the entrances of the main station and the Wannseebahnhof.

Apart from the amount of walking required, having three fully outfitted stations instead of one made the whole thing very expensive to operate: everything had to be done and paid for in triplicate. For that reason, the Potsdamer Bahnhof always ran at a considerable loss.

Yet another terminus (of sorts) was added to the complex little over a decade later, its only part still functioning today. On 15 February 1902, Germany's first underground transit service started to run from what is now Ernst-Reuter-Platz, via Zoologischer Garten station to the Potsdamer Bahnhof, where it ended. Today this section is a part of U-Bahn line U2. Only a part of the line was built underground, however: most of it ran on an elevated railway, but the connection to the Potsdamer Bahnhof was only granted on condition that it be built underground. As the underground (U-Bahn) system grew, it increased the Potsdamer Bahnhof's importance as a transit hub, further increasing pressure on the station.

Around the Potsdamer

Of all the Berlin stations, the Potsdamer probably experienced the largest shifts in its surroundings. When the first station was built, travelers leaving it would still be faced with the old city walls, but by the time the new one came around, these were gone and Berlin had started to assert itself as the confident capital of a new empire. As a consequence, the area surrounding what was now the Potsdamer Platz, Königgrätzer Straße (today's Stresemannstraße) and the nearby Wilhelmstraße developed into an upper-class neighborhood, adorned with diplomatic villas, ministries, embassies, museums, upscale restaurants, and other places of entertainment.

Yet some odd remnants of the past remained. The most eye-catching of these was a graveyard, the Dreifaltigkeitsfriedhof (Trinity Cemetery), directly in front of the station. Having a graveyard in front of the station is never a reassuring sight for a prospective traveler, of course, but the cemetery offered good opportunities for prostitutes to avoid the vice squad by posing as grieving widows, which contributed perhaps most to its removal in 1922. Plans to redevelop the cemetery into an extension for the station came to nothing, and it was eventually just turned into a smallish green on the square.

Within short walking distance, one could reach the Wilhelmstraße, the political center of the German Empire and arguably the most important power axis of the world around 1900. With the nearby Anhalter Bahnhof, the Potsdamer and its environs became a venue for diplomatic get-togethers and sessions of political gossiping. Receiving heads of state might have taken place at the Anhalter Bahnhof half a kilometer down the street, but the Potsdamer was the political station of the capital, where the civil servants and diplomats arrived from their homes in the suburb, and where the Emperor got into his imperial train.

The Roaring Twenties

Everything changed after 1918, as Germany's loss in the First World War made itself felt, with diplomatic prominence replaced first with revolutionary uproar and then with cultural exuberance. The communist revolution attempt of 1919 saw army divisions installed in many of Berlin's stations, but the sight of machine guns being set up at the Potsdamer Bahnhof, in the heart of Germany's government quarter, was perhaps more unsettling than any other.

As German politics slowed down to something at least relatively more tranquil during the 1920s and something almost resembling normality came about, the neighborhood gained yet another shift in identity. Next to the Potsdamer Bahnhof stood Kempinski's, renamed the Haus Vaterland during the war, which became

Fig. 2.11: Photos of the front vestibule at the Potsdamer Bahnhof, towards the train shed (left) and towards the street (right), c. 1929.

one of Berlin's main entertainment venues during the "gay twenties." Already a largish building in its own right, its bright lights totally dominated the street and the Potsdamer Platz at night, and helped to give Berlin its dynamic image in the years preceding Nazi dominance.

Of course, the Nazi era and then the Second World War changed all that once again: Kempinski's glory days were soon over after Nazi repression started. In its wake, the whole neighborhood lost some of its luster, and so did the station. The Ringbahnhof and Wannseebahnhof lost their function in 1939, when a subterranean S-Bahn station made them redundant. However, the diminished allure of the station did not prevent the Nazis from exploiting the prominently placed building for propaganda purposes. As a square-facing terminus in the middle of the city, it was a perfect billboard after all. When the Saar region was returned to Germany following a plebiscite in 1935, Berlin's representatives were jubilantly hailed at the station, whose front showed a line from a patriotic song from 1920: "The Saar is German" ("Deutsch ist die Saar"). Later, at the beginning of the Second World War, the slogan "Wheels roll for the war" ("Räder müssen rollen für den Sieg") was installed and expressed how the railways were supposed to play an important (and particularly sinister) role in the conflict.

Nonetheless, over time the Nazis planned to get rid of the Potsdamer Bahnhof. Had Albert Speer's megalomaniacal plans for the new capital (and "world capital") Germania come to fruition, the entire complex of the (and then some) would have been razed to make place for the "North-South Axis," a huge avenue leading from a new South Station to the "Hall of the People". We will return to those plans later.

The End, an Intermezzo, and a New Beginning

Fig. 2.12: Workers removing an unexploded shell from the ruins of the Potsdamer Bahnhof, 1945.

The razing took place anyhow, just not the way the Nazis had intended. By the end of the war, most of the square, street, and stations were gone. Of all the Berlin termini, the Potsdamer probably suffered the worst; not only was it bombed multiple times during the war, but the presence of Hitler's Führerbunker within spitting distance also guaranteed it a central position during the final days of the Battle for Berlin. Burned, bombed, and shot to bits, it was the among the few of Berlin's termini that could not be used again after the war and it did not even need to be blown up like some of its siblings. The Ringbahnhof was briefly used for S-Bahn services in 1945 and 1946 before it too was closed for good.

Because it was situated in the Mitte district (although protruding into Kreuzberg) the station area mostly became part of East Berlin after the war, but since all the tracks were located in the western part of the city few initiatives were undertaken to revive the building. What remained of the Potsdamer Bahnhof had been leveled by 1960. One year later, the Berlin Wall was constructed, and for years the area remained little more than a flat wasteland in front of the wall. In 1970, the inconveniently situated strip of land was given to West Berlin as part of a land exchange, but nothing happened (apart from the demolition of the remaining ruin of the Haus Vaterland).

After the Wall was removed in the early 1990s, yet another era in the history of the neighborhood and the station began. As part of the new city development in the area around the Potsdamer Platz, another railway station was built, this time below the surface. Above ground, an empty grass area clearly delineates the contours of where the old structure, Germany's very first station, once stood. Below, trains ride once again, just as they first did in almost two centuries ago.

Fig. 2.13: Potsdamer Platz today, with the entrance to the underground railway station in the foreground and the contours of the previous station still clearly visible.

3 The Anhalter: Grand Not So Central

When you walk up the stairs to the platforms, you become a traveler, and you're no longer in Berlin. Munich, Switzerland, Italy, the whole of the south draws you up the gray steps. [. . .]

The Anhalter is a romantic station, one for dreamers. The platform ticket costs 10 pfennigs. For that, you can walk the whole time along the platform and marvel at the large sleeping cars with their lowered shutters. The signs with the names of far-away stations are like identity cards for those who sleep behind them. – *Heinz Berggruen, "Bahnhofsgedanken" (1935)*

Fig. 3.1: Snapshot showing the platforms at the rear of the Anhalter Bahnhof, 1930s. Colorized photograph.

When you find someone waxing nostalgic about one of the lost stations of Berlin, chances are it will be this one. The Anhalter Bahnhof was famed for being Berlin's portal to far-away destinations in Austria, Hungary, Italy, and beyond. It looked the part, too, conceived in grand style in the late 1870s, when Germany had recently unified, Prussia had just given the French a damn good thrashing, and confidence rode high.

https://doi.org/10.1515/9783111381879-003

Compared to its brethren, and in spite of its reputation, the Anhalter's history is not particularly exceptional. It was not a fascinating conglomerate of ad-hoc decisions like the Potsdamer, fundamentally redeveloped in the same way the Stettiner was, a receptacle for huge streams of immigrants like the Schlesischer, or the center of a turbulent neighborhood like the Küstriner. After its opening, it functioned pretty well for 70 years, but became obsolete because of political reasons and was finally demolished. The story of that demolition, and subsequent efforts to resurrect it, figure among the most fascinating aspects of its history.

Fig. 3.2: Aerial photograph of the Anhalter (top) and Potsdamer (bottom left) stations, c. 1930.

In a way, it is a bit disingenuous to discuss the Potsdamer Bahnhof and the Anhalter Bahnhof separately; both were really part of the same, vast rail yard to the south of the old, walled city. But their reputations warrant it and Berliners of the time did not consider them to be a single station. The Anhalter's reputation has led many to assume that it was Berlin's biggest and busiest station – yet it was neither; the Schlesischer Bahnhof was larger, while virtually all the stations on the Stadtbahn saw more passengers in terms of absolute numbers. But most of them were commuters; things look slightly different if we consider long-distance travelers. Taking their numbers for 1894, the Anhalter ends up in second place after the Stettiner Bahnhof.[8]

Fig. 3.3: State occasions at the Anhalter: Emperor Wilhelm II receiving the Italian King Umberto I at the Anhalter Bahnhof, 1889.

The Gateway to the South

What it had going for it, however, were two important things. First, its ostentatiousness and size – always important in Berlin, a city that covets anything in-your-face. On its opening in 1880, the railway company proudly claimed that the station possessed the third-biggest hall in the world, after London's St. Pancras and Kansas City's Union Station.

The Anhalter's second strong suit was the *type* of destinations it served. In a time where the train was the only available means of efficient long-distance

---

8 Although it needs to be added that the Stettiner's traffic was highly seasonal (See chapter four).

Fig. 3.4: View through the Anhalter Gate towards the first Anhalter Station. Oil painting by Carl Daniel Freydank, 1841.

travel, the Anhalter Bahnhof offered the promise of whiling away at the Côte d'Azur, seeing the art of Florence, and visiting a sun-drenched coastal resort in Croatia. For most Berliners this kind of travel was entirely unattainable, of course, but the promise was a strong one and guaranteed the station an almost mythical status; it is a reputation that endures even today, having been fed by a century and a half of literature, music, and even cinema.

But let us go back a step. Before the grand Anhalter there was an earlier, much smaller one; when the Potsdamer Bahnhof was opened as Prussia's first railway station in September of 1838, work on the line between the capital and the state of Anhalt was already well under way. For its terminus, the Berlin-Anhalt Railway Company selected a spot close to the Potsdamer, about half a kilometer to the southeast on the Hirschelstraße.[9] At the time, this street consisted of two streets running parallel, with the old tariff wall between them.

The plot itself might have come cheap, but it did mean that the railway line itself diagonally traversed the western part of what was called "das Große Feld" (the large field), an exercise area to the south of the city. Prussia was not a coun-

---

9 This street was later renamed Königgrätzer Straße, after the victorious battle against the Austrians in the war of 1866.

try in which the interests of the army could simply be ignored, and the price was a heavy one: the construction of a bridge for the army across the Landwehrkanal and the purchase of a new exercise area near the Kreuzberg.

On the other side of the station, the situation was not ideal either: the Berlin tax wall was still in place, and without a gate opposite the new station. The railway had to construct its own one (the Anhalter Gate, fig. 3.4) to allow for easy travel to the station, with a new connecting street to the Wilhelmstraße. Again, this proved expensive, as the owners of the plots of land in-between had to be bought off and so the original plans for the station needed to be toned down considerably. To give the station a properly monumental "shop window," a square (later named the Askanischer Platz) was placed between the somewhat grimy Hirschelstraße and the building. However, much of the effect was lost a few years later, when the street-level railway that carried coal between the stations, the Verbindungsbahn, had to make its way across the street and over the square.

By 1840 most of the practical issues had been resolved, and the railway line could be opened on September 1. The new building had been built in roughly the same squarish style as the first Potsdamer Bahnhof. Much had been learned in the two years since the opening of the first railway, and functionally it was a marked improvement over its counterpart. Rather than terminating next to the station, trains now arrived at its back, for instance. However, it also shared many of its problems; most worryingly, it soon also turned out to be far too small.

The problem was compounded by the addition of a second railway connection granted to the Berlin to Anhalt Railway Company, this time to Dresden. For travelers it proved an enormous improvement, as originally the journey to Saxony's capital had gone via Leipzig, taking over 12 hours. In 1848 Prussia and Saxony decided to construct a less circuitous route which, when it opened in October of that year, halved the travel time. The connection proved popular but put even more strain on an already over-burdened station.

However, it was the construction of an even more direct route to Dresden (via Elsterwerda) in the early 1870s that made the necessity for entirely new facilities pressing. The new Berlin to Dresden Railway company that was set up to exploit the line insisted on its own station; an attractive location in front of the Belle-Alliance-Platz at the end of the Friedrichstraße (today's Mehringplatz) was considered but proved to be too costly. This meant that a new location had to be found somewhere in the sizable railyard behind the Anhalter: not just for a passenger station, but also for facilities, workshops, and a goods station. As the whole area was going to be redeveloped anyway, this presented a unique opportunity to also rebuild the Anhalter station itself.

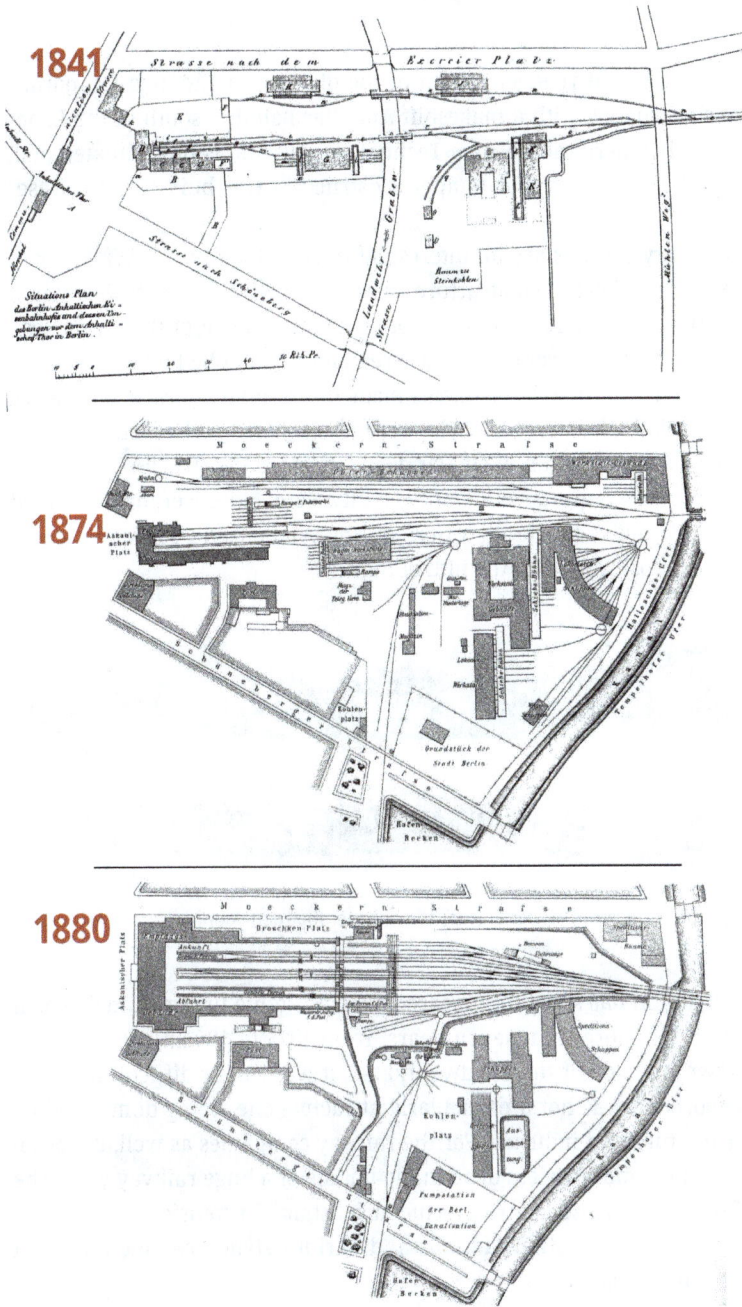

Fig. 3.5: The Anhalter's first 40 years. To scale.

The Forgotten Station

To make way for a grand redevelopment of the old station site, two temporary stations were constructed, with a makeshift Anhalter Bahnhof south of the Landwehrkanal, roughly where the German Technology Museum is now situated and, immediately to the west, a similar temporary structure was built for the Dresdener Bahnhof.

These temporary stations are an interesting story in themselves. We know of at least six instances where such structures were built, which were typically in a half-timbered style. This offered a number of advantages, some of them practical: the combination of wooden beams and plastered panels could be built quickly and cheaply, while the use of modular elements made disassembly, storage, and re-use somewhere else relatively easy. But there were also stylistic and artistic benefits. For starters, the visual repetition of similar elements made the structures look longer and therefore more impressive than they really were. Moreover, half-timbered buildings were regarded to be uniquely German and symbolized a sort of rural utopia that contrasted positively with the filth of the city.

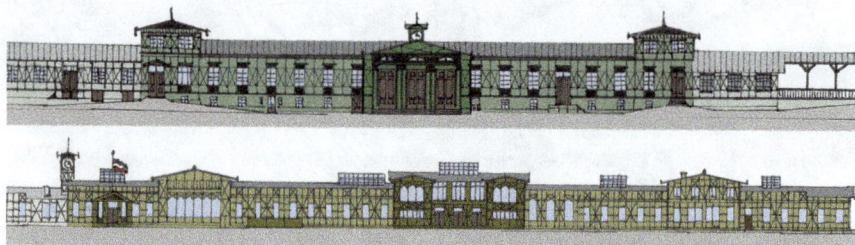

Fig. 3.6: Façade of the temporary Dresdener (top) and Anhalter (bottom) Bahnhof.

The Dresdener station has since disappeared, which is quite understandable as it did not function for more than seven years, from 1875 to 1882. Unfortunately, being on the "wrong" side of the Landwehrkanal, it was rather difficult to get to, and the neighborhood was not a particularly attractive one, being dominated by railway infrastructure and industry. For the railway companies as well, it was located inconveniently, more or less dead in the middle of a huge railway yard. The proximity of so much infrastructure precluded substantial extension without getting in the way of the two other stations and various structures such as train sheds and goods terminals.

The nationalisation of the line in 1882 led to the decision to transfer all passenger trains on the Dresden line to the new Anhalter Bahnhof, and the Dresdener therefore never received the more permanent building that had already been planned. After its closure, the temporary structure was gradually taken down. There is reason to believe that at least part of it was re-used at another site, possibly the Auswandererbahnhof (Emigration Station) at Ruhleben (See chapter ten).

Grand Central Conceived

In 1871, the architect Franz Schwechten began work on a new terminus for the Anhalt railway. While his first design already shows something we may identify with what was later built, there is also a very clear, earlier inspiration throughout the project: Karl Friedrich Schinkel and David Gilly's design from 1798 for a cathedral, which they in turn had based on an earlier, sixteenth century drawing by the French architect Philibert de l'Orme. Clearly, Schwechten was concerned with creating a true "railway cathedral." Moreover, particularly Schinkel was regarded with great admiration as the leading Prussian architect of the early 19th century, and the integration of his ideas granted Schwechten's own work greater authority. Although Schwechten's designs for the Anhalter lost some of their Schinkelian appearance over time, his influence would remain unmistakeable in the final product.

The third stage design of 1872 presents something of an anomaly; it contained two towers and three circular windows, rather than the repetitive gallery typical of the other plans. Probably drawn up by Schwechten's collaborators Orth and Knoblauch, it would serve to inspire the new Stettiner Bahnhof, a building completed four years before the Anhalter would be (see chapter four).

By late 1872, the final overall form of the station had been established, although a lot of work was still needed on the details. The terminal was more generously proportioned than any other station in Berlin (although a new Schlesischer Bahnhof would rival it within two years) with a width of well over 60 meters or, as newspapers did not fail to point out at the time, slightly wider than Unter den Linden, Berlin's central avenue. This made the new Anhalter the first station to deviate from the *Berliner Breite* (Berlin Width) of 37,66 meters, a long obsolete standard which had hindered the construction of sufficiently large buildings up to that point. With a length of almost 200 meters it was also longer than any other station in the city and the largest on the European continent.

Fig. 3.7: Anhalter Bahnhof, second design phase (Stufe C), 1872.

However, it ended up not quite as expensive as Schwechten had envisioned. Construction was started in September of 1875 after the terrain had been cleared of the remains of the old station, and services moved to their temporary facilities. One of the demands for the station had been that the tracks be moved above street level, which turned out to be exorbitantly costly because it meant building new bridges as well. Moreover, in the mid-1870s, Germany was in the middle of an economic crisis, and the railway company was forced to halt construction for more than a year from October 1875 due to financial shortages.

When the work started again in November 1876, Schwechten had been forced to simplify his design in order to save costs. Although the overall dimensions of the train shed remained intact, both side wings had become noticeably shorter and narrower, which meant that the waiting rooms and restaurants lost some of their opulence. Even the royal waiting rooms suffered, although the ramp for the emperor's coaches could be kept.

This crisis was one of the reasons why Bismarck forced through the nationalization of the Berlin-Anhaltische Eisenbahngesellschaft in 1882: he considered the railways to be too important to leave in the hands of market forces.

Stations as Architecture

Along with the slightly earlier Stettiner Bahnhof, the new Anhalter represented architectural innovation, and to see why we need to briefly delve into the history

Fig. 3.8: Anhalter Bahnhof, 1880.

of Berlin's other stations. The unveiling of the Ostbahnhof (popularly known as Küstriner Bahnhof, and not to be confused with the current Ostbahnhof) in 1867 had provoked some particularly harsh criticism; architects complained that the building in front of the train shed reflected nothing of the structure behind it, and hid something the Prussians were beginning to be proud of.

A more practical issue was that for the traveler it was useless: the entrance and exit were to the side of the shed, and the front building only contained administrative services and housing for railway employees, which also meant that a huge potential benefit of a terminal station – the concentration of services at the head of the platform – was left unused. The new Potsdamer Bahnhof of 1872 had only partly addressed these issues; while its front building was at least partly usable (but only as an exit, not as an entrance), it still posed a stylistic anomaly with the train shed behind, as much as the Küstriner did. This might have been all right two decades earlier but in the meantime architects had started to integrate a building's function into its design.

The controversy was given greater weight because around this time, the German profession of the Baumeister – a blend of contractor, engineer, and architect – was beginning to be diversified into its construction and design/aesthetic parts. Architects of a more modern kind began emancipating their profession and were no longer content with the utilitarian setup of previous railway architecture. Moreover, as Berlin was becoming an imperial capital and a growing metropolis, its planners looked to examples from outside, especially in London and Paris. Notably, Baron Haussmann's redevelopment of the Parisian inner city utilized the French capital's large railway termini as imposing bookends to its new, large boulevards – something which contrasted markedly with the way in which many of Berlin's stations had been tucked away.

Grand Central

Fig. 3.9: Hermann Rückwardt, Anhalter Bahnhof, train shed, 1880.

The new building was ceremoniously opened by Reich Chancellor Bismarck on 15 June, 1880. At the Askanischer Platz in front of the station, a crowd of hundreds had gathered to take a look at the new station and be among the first to enter it at 3.30 a.m. The first train, the 5.40 a.m. to Lichterfelde, had to be extended several times to accommodate the 700 people that wanted to be on it, and even then it could not leave because almost 200 railway employees wanted to tag along as well. The 58-axle express eventually departed with a five-minute delay.

Fig. 3.10: Anhalter Bahnhof, entrance hall (left) and royal waiting rooms, 1880.

From the outset, the station was meant to impress. The opening doorway led into an opulent grand hall, which contained the vestibule and ticket offices. To get to the trains, passengers had to ascend a huge staircase, at the top of which the even more impressive train shed revealed itself. Architectural reviews were favorable from the outset, both within and outside Germany. The *Leipziger Illustrirte Zeitung* declared: "Here we do not see that sickly compromise between utility and monumentality, which often shows itself to such a troublesome degree in railway buildings; rather, the building is fully and magnificently conceived and masterfully executed in all its aspects."

Like the Stettiner four years earlier, the Anhalter did away with the traditional division between arriving and departing passengers, and so became much more convenient for those wanting to switch trains. Upon arrival, travelers could leave either by the side or the front, but there was no longer a general entrance building to the right (west) of the station; that space had been allocated to waiting rooms (this was still the side where trains left) and restaurants, but could only be reached from within the station. The outside could therefore be used for logistical operations, such as goods and baggage handling.

The royal rooms were situated on the eastern side of the building. While these obviously served the members of Germany's royal families (the *Allerhöchste Herrschaften*), they also functioned as something of a showcase for the ability of

Fig. 3.11: Anhalter Bahnhof, First class waiting room (situated in the northwest corner of the hall).

the architect and the prestige of the railway company and destination. Of all the royal rooms in Berlin's stations, these were among the most used since this was where visiting heads of states typically reposed after their arrival.

With a better station came a better neighborhood. Gone were the days, when the street in front of the station was being dominated by the fumes of the Verbindungsbahn; the Königgrätzer Straße, as it had been renamed, had become a place where upper-class hotels catered to the beau monde that had just arrived or was waiting to depart; the Excelsior, Askanischer Hof and Habsburger Hof were among the best in town, with prices to match.

From now on, this was the place where foreign dignitaries arriving from the south were received in the capital and where the wealthy set out on their vacation. Trains departed for Dresden, Magdeburg, and Frankfurt, but also for far more exotic destinations such as Istanbul, Athens, the Mediterranean coast, or Rome. Still, the station also remained an important hub for local and regional traffic.

That is not to say that despite its generous dimensions and its six platforms it did not immediately suffer from the same, continual growing pains that afflicted all the Berlin stations. This became immediately apparent when, two years after its opening, Dresden services were diverted to the Anhalter from the ill-fated Dresdener Bahnhof. Quite big still was not large enough, as it turned out. Over the years, additional tracks were laid – most of which ended outside the big train shed – facilities added, and platforms extended. Although more extensive remodelings were conceived, these were never realized, not least because of the First World War and the subsequent economic crisis of the 1920s.

A much-publicized addition took place in 1928, when a tunnel to the luxurious Excelsior hotel was installed in the basement, allowing its guests to move to and from the station without having to brave the elements and Berlin's hectic traffic on the Königgrätzer Straße. Some stores were even opened in the tunnel, but as it turned out the main attraction was its novelty; few people actually used the tunnel and the shops weren't able to survive.

Wrong Place, Wrong Time

From the turn of the twentieth century onward, it became increasingly clear that in the long run the Anhalter was on the way out. Plans to rationalize the city's railway infrastructure came to focus on the creation of a north-south tunnel between the area around the Lehrter in the north and the Potsdamer Bahnhof in the south – the situation that exists today, in fact. The Anhalter, for all its splendor, was just slightly in the wrong place; a U-Bahn connection could never be realized, and it took until 1939 for an underground S-Bahn link to be opened. Even then, the connection to the Anhalter Bahnhof required constructing a somewhat unseemly curve in an otherwise mostly straight line.

Later plans for the reconstruction of the railways in Berlin mostly ignored the "Gateway to the South." However, the station building's reputation might have saved it regardless. Even Hitler and Speer's megalomaniacal plan for a new *Welthauptstadt Germania* from 1938, which didn't shy away from tearing down entire swathes of the city, still saw a purpose for the building as a museum or a swimming pool.

As we see with depressing regularity, it was the Second World War and its consequences that hailed the end of Berlin's grand "gateway to the south." From 1933 onwards, it was where those that feared repression by the Nazi regime went into exile; Heinz Berggruen, a Jewish art collector quoted at the start of this chapter, was among them. Things took a turn for the even more sinister when, from June 1942 onwards, the station was used to transport the city's elderly Jews to Theresienstadt concentration camp. This was done using regular trains, unlike at the more out-of-the-way Grunewald station from where Jews were transported in the cattle carriages that we have come to know (see chapter ten). The Anhalter was still in the middle of the city of course, and used by other travelers, so too much brutality might have been a tad too conspicuous.

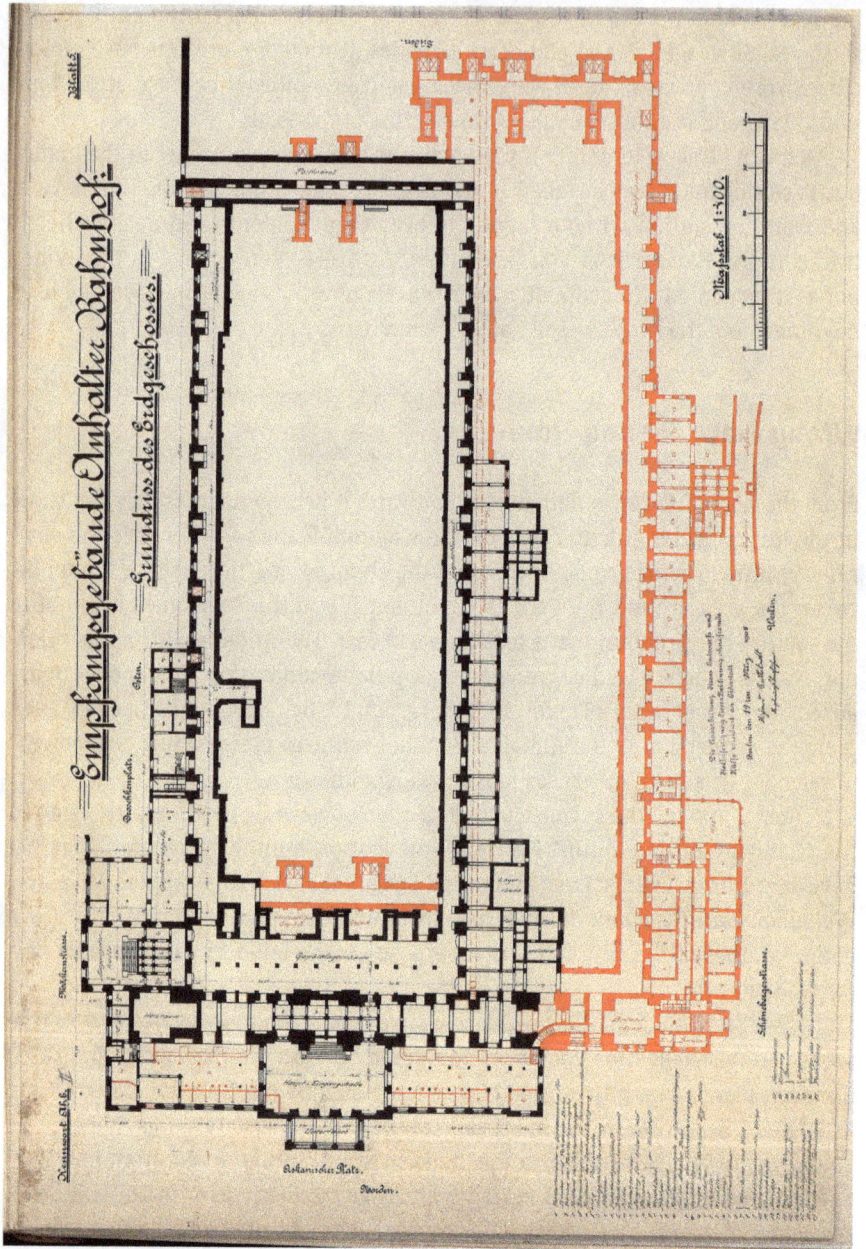

Fig. 3.12: Richard Johann Guttstadt (1879–1942), Plan for the extension and renovation of the Anhalter Bahnhof, Berlin. Schinkelwettbewerb 1908.

Fig. 3.13: The Anhalter Bahnhof in ruins, mid-1945.

Bombs heralded the beginning of the end of the mighty Anhalter. The building was heavily hit on the night of February 3, 1945, with the rear part of its roof not surviving the bombs and the subsequent fire. The damage could be cleared up to some extent, but train traffic ended in April of that year during the Battle for Berlin. Citizens of the heavily damaged city took refuge behind its monumental walls, as well as in its basement and the Excelsior tunnel. On April 29, the Russians attacked in full force, but around 10,000 people were evacuated through the S-Bahn tunnel to Stettin station.

Bombed and shot to pieces, the station building was nonetheless reactivated after the end of hostilities, although the remaining part of the roof had to be removed for safety reasons. Train services, mostly to the outskirts of Berlin and further on into the Soviet Zones, were resumed in August 1945, and continued, with an interruption during the Berlin Blockade of 1948, until 1951. In that year East German citizens were formally prohibited from visiting West Berlin, which left the Anhalter – a West Berlin station whose tracks reached into the GDR and East Berlin – without passengers, leading to its inevitable closure. The "Gateway to the south" had ceased having a south to be a gateway to.

That Entrance

Fig. 3.14: Portico of the Anhalter Bahnhof, 2024.

The Anhalter's almost-disappearance is a typical story of damaged structures in old Berlin. There were a number of factors that contributed to its end, though. Firstly, as we saw, the Anhalter had officially ceased to be a viable station for destinations in East Germany after 1951. Several plans were made for a new Anhalter Bahnhof, though, even after political circumstances had made it obsolete. A first design was presented at the Constructa Building Fair in Hannover as early as 1951, but more plans followed in subsequent years. However, these never went beyond a sketch design.

The second contributing factor was the craving in West Berlin for any modern thing not related to an ignominious past. This was the time when the authorities there provided subsidies to have the old ornamentation removed from housing blocks and so give the city a more contemporary appearance, which meant that buildings such as the Anhalter Station could count on little love from the authorities. Finally and perhaps most effectively, at that time West Berlin was not exactly known as a haven of bureaucratic integrity. Long story short was that someone got a good deal for the bricks, so the building came down in 1960, despite vocal opposition. Ostensibly, the tile work on the outside was to be re-used for reconstruction housing, but since everything had been built by Prussian engineers, the tiles could not easily be removed from the bricks, nor the bricks from one another, and all were eventually thrown away.

However, just before the demolition work was done someone suggested the portico be spared as a reminder of the horrors of the Second World War. This is where it still stands, amid as desolate an urban landscape as you are likely to encounter in Berlin. Plans to build an entirely new station came to nothing for obvious reasons. The area once occupied by the train shed is now a football pitch, bookended by a concert hall, the Tempodrom. Behind that, however, one can still find the platforms that were once part of the Anhalter – a far more obscure memorial, now overgrown and often the home of the capital's homeless community.

In a way, the fate of the Anhalter contributed to something of a turnaround in attitudes towards Berlin's architectural heritage. The ferocity of the protests was such that not only was the demolition not final, but it also caused the authorities, at long last, to rethink the way in which they handled such cases. This is what likely helped the Kunstgewerbemuseum (today's Martin-Gropius-Bau) to survive a few years later.

Museums and Reconstructions

From time to time, calls for the Anhalter's reconstruction have appeared and gone away after a while. There is no longer any use for a station on the site, but because it has not been built over so far hope still remains of rebuilding the glory days. But what is one to do with such a huge hall? One suggestion has been to use it as an entertainment venue like the Tempodrom, or – multiple times – as a vast swimming pool.

Perhaps, a stronger argument against reconstruction may be the present situation of the surrounding area. Almost entirely flattened in the final days of the war, it was later redeveloped in the least inspiring 1960s architectural style and has effectively been ruined. A rebuilt Anhalter would look entirely out of place if its surroundings did not receive some serious attention as well. Luckily, it's still – somewhat – possible to take a look inside: in 2022, Germany's Museum of Technology launched a website containing a virtual reconstruction of the station, along with a virtual tour around its remains in Berlin, where people may explore the station at their leisure using their phone.[10]

---

10 Visit https://anhalter.technikmuseum.berlin/.

For now, the portal's most likely future seems to be as the entrance of a new Exile Museum that is planned on the site. Again, this has raised protests, not so much because of the museum itself but rather because its construction would likely permanently prevent a reconstruction of the station. While the idea has gained political support, museum professionals have also been critical of the elitist concept behind such a museum. After all, it would be mainly dedicated to more famous people that fled the Nazis and who wrote about their experiences, not so much the many ordinary Germans that failed to do so. At the time of writing, the founding of the museum seems to have arrived at an impasse, mostly because of a failure of its initiators to secure sufficient funding and political support hasn't translated into money – yet. Be it out of shame for its casual destruction or some other reason, a hesitancy remains about replacing what was once Berlin's most prized railway cathedral with something else.

4 Stettiner Bahnhof: Vacation Station

Fig. 4.1: Horse-drawn carriage in front of the Stettiner Bahnhof, 1898. Photo by Waldemar Titzenthaler.

Above me, the station clock flickers in the sun. It's been broken for a year. Nobody knows why it hasn't been repaired yet.

A fat, red-cheeked man, travel bag in hand, struts across while looking around searchingly. Already, a lady who, like me, stands around on the corner to earn money, is helping him. She calls him "Uncle" and "Sweetheart" so that he is completely moved and confides in her leadership. The other girls are annoyed that Emma has caught herself yet another one. "Man, look at her with that good-for-nothin'". – *Hardy Worm, 1921*

Until the end of the Second World War, the Stettiner Bahnhof, or Stettin Station, was also known as Vacation Station, the *Urlaubsbahnhof*. From here, Berliners escaped the capital's sweltering summer heat by train, in order to cool off at the trendy resorts along the Baltic coast: Hiddensee, Rügen, and Usedom, the places where German tourism was to a large extent invented. The Kaiser himself liked to visit the island of Hiddensee, and his people followed in his wake; in the early

https://doi.org/10.1515/9783111381879-004

summer, they would escort their families to the coast. In terms of passenger numbers, it was Berlin's busiest.

In spring, summer, and early autumns, the station square in front of the side of the station was continually abuzz with *droschken* (small coaches) and porters, getting the vacation-goers and their monumental luggage into and out of the trains. Up to the Second World War, no station in Germany handled more suitcases and trunks than the Stettiner. The family might stay on the Baltic coast for as much as four or five weeks, but the husbands habitually returned to the capital to work during part of the week.

Straw Hats

Fig. 4.2: "Straw Hat Widowers".

Of course, they did not only work. Being liberated of responsibility for their next of kin, they turned into Strohhutwitwer, "straw-hat widowers," looking for entertainment of various kinds as soon as they left the station; they visited the bars and cabarets in the Poet's Quarter, the Beer gardens near the Kastanienallee and Rosenthaler Platz, and many of them ended up taking advantage of the one-hour hotel rooms in the neighborhood with their new-found "wives" – checking in under a variety of fictitious names. Sometimes these were surprisingly inventive, but most of the time, a common "Schmidt" or "Müller" would do nicely. Others made a deal with the porter of a Mietskaserne (tenement), who in exchange for a tip would turn a blind eye while one of the one-room apartments was being used for business. The area around Stettin Station was not the sort of neighborhood where you left your children on their own, or where you would like to be spotted by an acquaintance.

It was also predominantly a working-class quarter, where various proletarian movements enjoyed a very strong following. During the early 1930s, one of the Nazis' greatest triumphs was being able to turn this neighborhood around to their side; from then on until the end of the war, it remained staunchly National Socialist. Today, it has been mostly gentrified, with only the Airbnb-infested Torstraße an unwitting reminder of its less-than-stellar past.

The railway station building itself always looked a bit dusty too, almost as though the reputation of the neighborhood had brushed off on it. But the more concrete reason was that the northern part of the city was also the home of many factories, most notably the Borsig locomotive works in Tegel, with the smog quickly settling on the building. It also meant that the mood in the neighborhood was firmly proletarian, with scores of factory workers using the smaller Vorort-bahnhof (suburban station) to reach their workplace. Taken together, it definitely created a very different atmosphere than at the elegant Anhalter or the teeming Potsdamer stations, with Hardy Worm's observation of the broken clock adding to the overall impression that, from an early point, the Stettiner made a slightly dilapidated impression.

But as we saw, it was not entirely negative; for Berliners, the station marked the beginning (and end) of their Baltic holidays. In winter, however, the Stettiner largely fell into hibernation mode. The factory workers would still come in of course, but they mostly used the smallish Vorortbahnhof on the side.

First Station

The first attempts to open a line between Berlin and Prussia's biggest port at Stettin (today Szczecin in Poland) started in the late 1830s. On the Berlin end, planners immediately ran into problems; where the south of Berlin consisted mostly of flat terrain, the north was much more hilly. Plans to erect the new terminus in front of the Rosenthaler gate in the Berlin tax wall had to be canceled because it required the construction of a prohibitively expensive tunnel through the Weinberg, a hill to the north. However, an alternative site was found in the former gallows' field on the Invalidenstraße, more or less in front of the Hamburg Gate on the Torstraße.[11]

The Stettin-Berlin Railway company opened its first station on 1 August 1842, in the wake of the Potsdamer and Anhalter stations. In the first months, trains only ran up to Eberswalde, but from mid-1843 onwards Stettin was reached. In terms of design, the station building differed somewhat from the Anhalter and Potsdamer, as it seems to have used noble villas rather than office buildings as its main source of stylistic inspiration; its neoclassical leanings were therefore increased and can still be seen on other stations along the same line, such as those in Bernau and Eberswalde. The building also mimicked the shortcoming of some earlier stations, notably the lack of outside protection for travelers. However, it did offer them more luxurious facilities, such as separate waiting rooms for each class, inside the building.

Fig. 4.3: An impression of the 1842 Stettiner Bahnhof by Milo van de Pol.

---

11 The field was no longer in use after the introduction of the guillotine as the favored method of execution. The (smaller) execution site was moved to the Chausseestraße, more or less (and rather sinisterly) where the German secret service (BND) building now stands.

Like its predecessors, the station would see continuous extensions and additions to cope with the ever-growing stream of passengers and goods. However, its problems became even more acute when plans for the addition of a new railway line to the city of Stralsund on the Baltic began to take shape in the 1860s, and it was clear to everyone that the old building was in no way sufficient to handle even more traffic. The problem wasn't just the addition of a new line – it was the type of travel. Mass tourism would only come later, but people already began to use the railways for non-critical journeys: to go on day trips to the countryside, for instance, or to visit family around religious holidays, particularly Easter and Pentecost. The building was ill-equipped to handle this onslaught, and in the beginning of the 1870s work started on a new station.

A New Design

The development of the new Stettin station is connected with three other stations in three different towns: London's Fenchurch Street (1853), Paris's Gare de l'Est (1849) and Berlin's later Anhalter Bahnhof (1880). During the extensive design phases for the Anhalter, an alternative design was presented that differed markedly from earlier and later sketches. Drawn by August Orth and Edmund Knoblauch, it quite obviously took Paris's Gare de Strasbourg (today the Gare de l'Est) as its main inspiration.

This design differs noticeably from the others made for the Anhalter Bahnhof, but we can already see some elements that would end up as part of the new Stettiner: particularly its large, rounded form flanked by two tower-like structures. A second design from December of 1872, drawn by Franz Schwechten of Anhalter Bahnhof fame, shows something even more familiar. The three large, arched windows became part of the façade, which was now visibly an extension of the train shed. In both respects, Schwechten's design followed a trend set by the design for Fenchurch Street (1853), one of the smaller London termini. The shared DNA is clearly visible, although the Stettiner included many improvements and extensions when compared to the London example.

One was the train shed itself. The architects had learned from the example of two Berlin stations that had opened shortly before, the Ostbahnhof (or Küstriner Bahnhof) of 1867 and the Lehrter Bahnhof of 1871, both of which used a round barrel roof and made it possible to build a high, airy but also light structure. They had also taken notice of London's St. Pancras station (1868), which combined this method with the use of cast-iron girders, allowing for quick and relatively cheap construction.

Fig. 4.4: Gare de Strasbourg, Paris (top; today the Gare de l'Est) and the second design phase for the Stettiner Bahnhof by Stettiner Orth and Knoblauch.

Fig. 4.5: Second Stettiner Bahnhof, original appearance before 1896.

That Orth and Knoblauch could integrate Schwechten's design into their own work (they were also collaborating on the Anhalter around the same time) tells us something about the consensus that had arisen around the construction of a railway station. As a whole, Stettin Station was very much the result of iterative improvements on previous examples, coupled with contemporary aesthetic cues that favored a more overt expression of the building's purpose in its form. The end result, however, was quite revolutionary by Berlin standards, perhaps even more so than the more famous Anhalter Bahnhof.

Throughout the process, the railway companies played an important part. However, shortly after the building was finished, both railway companies that were going to use it were nationalized by the Prussian state. Nationalization could not prevent that within the building many services remained divided between those for the Baltic Nordbahn to Stralsund and those for the railway to Stettin, with separate ticket offices and personnel.

A New Beginning

The Stettiner represents an important new stage in the evolution from the "stationary umbrellas" to purpose-oriented travel lodges. Firstly, in terms of design, the shape of the hall made the structure immediately recognizable as a train station or, in more modern terms, an example of the "railway cathedral". Moreover, it had an entrance and exit (which were no longer separated) at the front; not only did this give the building a much more welcoming appearance, integrating it with the city outside, but it also made the most of the terminal platform and services inside.

Entering the hallway, people could directly continue into the train shed, or drop off their luggage. Most services were integrated; only purchasing tickets required passengers to either turn left for the Stettin line or right for the Baltic one. However, the station retained some concessions to existing practices; for one thing, the shed and reception hall was still built according to the *Berliner Breite* ("Berlin width") of 37,66 meters, even though the two lower buildings to the side extended the total width of the structure to over 60 meters. The Anhalter, opened slightly later, was the first station to finally do away with that restriction.

Still, Berliners embraced the new; it was much more *theirs* than the exotic but also somewhat inaccessible and pompous Anhalter Bahnhof. Walter Benjamin expressed how, for them, the station symbolized the "dune landscape of the Baltic Sea" which emerged "like a mirage, supported by the yellow sandy colors of the station building and opening those wide horizons behind its walls." In other words, it was where their holiday began, and where they could finally leave the stresses of life in the capital behind them. However, not everyone described the station in such favorable terms, or, as Alfred Kerr put it, "In general one feels most comfortable in Berlin at Stettin station, when one is about to leave it."

A rare, if moderate, voice of criticism could be read in the *Deutsche Bauzeitung*, an architectural journal. Karl Fritzsch's disapproval primarily concerned the execution of the building: "The powerful motifs used for the façade design, the harmony between the interior and exterior and the clear expression of the building's purpose in its appearance would be advantages that could secure the new reception building of the Stettin railway station the first place among all recent designs of this kind in Berlin, if the architectural development of these elements were at the appropriate level." He found it particularly "regrettable that the beautiful impression of the worked limestone has been spoiled by a coat of diluted cement (applied to protect it from the weather)." Indeed, this was something that contributed to the somewhat dirty appearance of the building.

Holidaymakers now wished to spend their summer weeks on the Baltic coast, and such structural tourism turned out to present a huge challenge because of its seasonal nature. To give some idea of the problem: in 1894, the Stettiner was Berlin's busiest single station by some distance (if we do not count the Stadtbahn as a whole), but most of those travelers only used the station during the spring and summer when the station again proved to be far too small.

There was another, major issue. The tracks leading to the station were all situated at street level, causing dangerous situations and countless traffic jams while everyone waited for the many trains to pass. From May to September, traffic was so intensive that those streets were blocked for much of the day, a problem which was worsened by the circuitous route the tracks had to take through Northern Berlin to reach the station. In addition, the terminus itself was much too small to handle all those people, especially since those holiday-goers also came with a huge amount of luggage.

Fig. 4.6: Eberswalder Straße (Nordbahnhof) station in the 1930s. The platform to the right was the one used for passenger trains during the 1890s.

The Stettiner was part of a larger complex of railways in the north of the city, together with the Nordbahnhof, a goods station to its east along Bernauer Straße (the location of today's Mauerpark). During the 1890s, the overcrowding of the station got so badly out of hand that it was decided to divert part of the passenger traffic to the goods station. A makeshift passenger terminal served here between 1892 and 1898.

By that time, it had become clear that the situation at the Invalidenstraße needed a more encompassing solution; a choice had to be made between building an entirely new station or significantly extending the existing one.

Pushing Up: The Renovation of 1903

Eventually, the railway authorities settled on doing both, and ordered the architect Carl Cornelius to redesign the station. The shell of the old station was left intact, but entirely revamped, and the entire body of tracks lifted to alleviate the worst traffic problems. The project was carried out in stages during six years: to begin with, a new station for suburban connections (the Vorortbahnhof) was constructed at the west of the main building and opened on May 1, 1898. Built in a style that alluded to that of the main station, it had the added advantage of creating a social separation: factory workers on their way to the Borsig plant no longer needed to mingle with families going on holiday. A temporary structure in the usual half-timbered style for holiday specials was opened on its east side just under a year later, which allowed for the reconstruction of the main station to begin in earnest.

While it might have looked more or less the same from the outside, the 1903 renovation created an almost entirely new building. The tracks were raised by about three and a half meters to well above street level, at last removing the constant traffic congestion issues. The station itself was extensively restructured, but great care was taken not to make the changes too conspicuous; for passengers, the most immediate change would have been that due to the raising of the tracks, they now needed to ascend further stairs to reach the platform level, just like they did at the Anhalter Bahnhof. Likewise, the roof of the hall was lower above the passengers' heads by the same three and a half meters, although it remained at an impressive height.

To the east of the old train shed, three smaller ones were constructed in a similar style, devised to keep the visual consistency of the complex as intact as possible. They helped to double the surface areas of the station, which not only benefited the tracks and platforms but also passengers, who could now enjoy spacious waiting rooms at the front of the new halls. Finally, a longish (179m) tunnel further north now allowed people to cross the vast rail yard without having to walk all the way around the front of the station.

After the reconstruction, the original train shed was exclusively used for arriving trains; the new, lower halls were to be used for departing traffic, including

Fig. 4.7: The temporary station at the Stettiner Bahnhof, around 1900.

Fig. 4.8: Hall of the Stettin station, 1896 (left) and after 1903 (right). The difference in height is clearly visible, especially in the central windows.

the specials to the Baltic coast. For at least the time being, this brought the previous capacity issues to an end; moreover, the authorities had learned from the continual shunting issues at other stations, and the number of sidings (tracks designed to park trains) was extended to 11.

Some problems remained nonetheless; the platforms were still awfully short for the long holiday express trains at a mere 325 meters, and the limited number of tracks feeding traffic into the station remained a bottleneck.

Despite these remaining issues, the station appears to have functioned pretty satisfactorily in spite of continued growth in passenger numbers to a massive 4,5 million in 1922. Significantly, when the architect Harald Roos proposed the restructuring of the Berlin railways in 1929, the Stettiner was the only one of the old termini to be left standing in his plans. Around 1930, 1,5 million passengers were handled annually and the station managed to cope quite well.

While it could hardly boast the same number of international connections as the Anhalter, the Stettiner Bahhof did connect to Sweden via the Sassnitz ferry on the island of Rügen. This proved to be a crucial connection in both world wars when Swedes were able to visit Germany – and conversely, Germans could abscond to neutral Sweden.

Fig. 4.9: The appearance around 1910 on a colorized postcard (compare Figure 4.5). These colors seem to be reasonably accurate if we compare them to color photos of the ruin after 1945.

The first decades of the twentieth century proved to be not particularly eventful in the history of the station, with the most significant change brought shortly before the outbreak of World War Two, with the opening of an underground connection between the Anhalter, Potsdamer, and Stettiner stations (the north-south tunnel serving today's S1, S2 and S25). An underground station (today Nordbahnhof station) replaced the old Vorortbahnhof, which was scheduled for demolition.

Nordbahnhof

Ironically, today that old suburb station is one of only a few parts of the complex to survive. Predictably, the Second World War brought great destruction to the station, but not quite as extensive as in other cases. It had already suffered bomb damage in 1943, which caused the 1876 hall to burn out; the newer halls survived mostly intact, however, and this is where train travel recommenced in July 1945.

Fig. 4.10: Paul Grunwaldt, *Refugees at the Stettiner Bahnhof,* Watercolor on cardboard, 1946.

Six years later, the complex, now in the GDR, was renamed the "Nordbahnhof"; Stettin was no longer in German territory, and it was deemed politically unacceptable to refer to once-German destinations. This was somewhat confusing since there already was a Nordbahnhof, the goods station mentioned previously, which was in turn renamed Eberswalder Straße station. For similar reasons, the Schlesische (Silesian) Bahnhof became "Ostbahnhof" at the same time, despite there already being a (now defunct) Ostbahnhof.

Only months later, the Stettiner also became functionally obsolete. The railroad to Stettin, now Polish Szczecin ceased to be the lifeline it once was, and while the tracks did also connect to the northern coast, they ran through West Berlin for some distance, with all the problems that entailed. After closure the unused complex lingered on, increasingly ramshackle, until it was finally torn down in 1958.

Remnants

Today, only two fragments of what was once Berlin's busiest single station remain. One is the old suburban station, which had already become defunct before the war; today, it looks a bit lost amidst the many anonymous high-rise office blocks in the vicinity. It is now being used as an event venue; in previous years there was a restaurant in the building, but that appears to be gone.

But at least it has been lovingly restored, which cannot be said of the Stettin Tunnel, a 179-meter underpass that served since 1896 as a pedestrian connection between the Gartenstraße and the Schwartzkopffstraße. It is open to the public, but only for special tours, and even then no one can traverse its entire length, since a major gas pipe traverses the eastern end of the tunnel these days.

The Stettin station has not disappeared entirely, however. The area remained in the hands of the East German Reichsbahn following the demolition of the station, and after Germany had reunified the Deutsche Bahn built the offices for its services division on the site (and gave it the somewhat tired title of "Nordbahnhof-Carré").

While stylistically quite different the new building was forced to conform to the contours of the old station site, so that even today passers-by may get an impression, even if only a faint one, of what was once Berlin's busiest railway station.

Fig. 4.11: The Stettiner Vorortbahnhof today (photo taken 2020).

5 Schlesischer Bahnhof: The One That Survived

Fig. 5.1: The Schlesischer Bahnhof from the west, late 1920s. Note the single remaining tower. Colorized photograph.

> Berlin was silent, soundlessly raising and lowering its dull body in its sleep. Only the Silesian Station released noise, whistles, hissing steam – and occasionally a few shadows of people who wafted through the underpass and dissolved somewhere in the darkness. – *Georg Fink (pseudonym of Kurt Münzer), Hast du dich verlaufen? (Did You Lose Your Way?, 1930)*

In this story, the Schlesischer Bahnhof (Silesian Station) is something of an exception because it is the only one of Berlin's great termini that is not only still around, but also serving its original purpose. Today called Ostbahnhof and converted from a terminus into a through station almost 150 years ago, it continues to form a crucial part of the capital's public transport infrastructure. During its existence, it has gained a reputation shaped by the explosion of the city around it, class struggle, two world wars, and a revolution, various autocrats, and a serial killer. Usually, that reputation wasn't stellar, which may explain some of the stepmotherly treatment the building has had to endure.

https://doi.org/10.1515/9783111381879-005

During the 1920s, the area around the Schlesischer Bahnhof, Berlin's "gateway to the east" was little better than the neighborhood near the Stettiner, however, it was dominated by sleaziness of a somewhat different character. Of course, there were the thieves, the prostitutes, and the swindlers going about their business inside the station, in the countless bars that encircled it, and often in the doorways of the Mietskasernen, the tenement buildings that were among the worst in the city. But these denizens of the underworld were hardly free to go about as they pleased, for this was the home turf of the Ringverein "Immertreu" ("Always Faithful"); criminal activity in western Friedrichshain, the city district that contained both the Schlesischer and Küstriner stations, was very much regulated and retribution would come swiftly to those who thought they could ignore it.

The Ringvereine were Berlin's flavor of the Mafia, criminal organizations that operated as a type of villains' union, ostensibly dedicated to the rehabilitation of convicts; they handled their reintegration into society, by providing their members with jobs when they got out of prison for example, as well as supporting their families. Then there was a different sort of activity, the sort that took place in the shadows, because the Ringvereine organized crime as well. What made the setup somewhat confusing to outsiders and law enforcement alike was that they also usually took their social responsibilities seriously; they were even known to force criminals onto the straight and narrow if their families needed them to.

In the middle of all this activity stood the Schlesischer Bahnhof, one of the city's major transport hubs, a point of arrival for the destitute from the east and a place that lived through more German history than any other in the city: one that emerged from all that time as the only one that lives to tell the tale. Terming it the "Schlesischer" is also a choice; probably no other single railway station has seen as many name changes. Starting out as the "Frankfurter Bahnhof" upon opening in 1842 and again in 1869, it was also known as the "Niederschlesisch-Märkischer Bahnhof", officially re-named the "Schlesischer Bahnhof" in 1882, then "Ostbahnhof" after the Second World War, "Hauptbahnhof" in 1986, and finally back to "Ostbahnhof" ten years later; the name it still carries today. Each of those name changes carried its own significance, be it logistical, organizational, or political.

In the Fields

The contrast of the situation in the 1920s with the circumstances at the time of the Schlesischer's beginnings could hardly have been greater. When the first station on the site was opened, the city buildings were still distant. The Schlesischer Bahnhof is also a special case among Berlin's termini because its history shows an uncommonly iterative approach; there were no grandiose demolitions and total rebuilds, but rather a continual, steady adaptation of existing structures into new ones – either intended or improvised. As a consequence, the attentive visitor to today's Ostbahnhof may still spot a few features that stem from the 1860s.

Fig. 5.2: Frankfurter Bahnhof, 1843. Note the lack of surrounding buildings.

The oldest Frankfurter station was among Berlin's first generation. Trains started rolling out of it on 23 October 1842, only weeks after they had first left the Stettiner, and the new station again demonstrated that lessons had been learned from the construction of the pioneering Potsdamer and Anhalter stations. It was by far the most advanced of its generation, but unfortunately it also created unhelpful precedents of its own.

The Frankfurter Bahnhof was the first Berlin station to separate arriving and departing passengers by assigning them their own platforms, even buildings. For this purpose, a reception building for departing passengers was constructed on the north side of the complex and an entirely separate arrivals building was built in the south, near to the river Spree; since most passengers would have arrived from the Frankfurter Allee in the north, this must have seemed like an obvious setup. Initially, it did not really cause issues, since the rails to Frankfurt were single-track, but once the track were doubled in 1847 it proved cumbersome, since the station was now built the wrong way around. Trains drove on the right in Prussia, and this meant that all trains entering and leaving the station needed to cross the tracks to the other side, which greatly complicated operations, all the more so because immediately after pulling out of the stations trains had to cross the Fruchtstraße, an increasingly busy road.[12] This crossing would continue to hamper the development of the Frankfurter Bahnhof over the coming decades.

Fig. 5.3: Entrance building of the Frankfurter Bahnhof (left), c. 1843. Copperplate by Loeilliot and Hintze, 1843.

Although passengers would still alight beneath a small canopy, the basic template of early Berlin termini, including many of the second generation, was established: administration building at the front, combined with separate entrance and exit buildings behind. It had also been proportioned rather more generously than previous stations but, like the Potsdamer, Anhalter, and Stettiner before it, the station was enhanced by additional structures and tracks more or less continuously after its opening.

While those stations needed to be built outside of the *Akzisemauer* (tax wall), the Frankfurter was unique in being the only station built within it. This was possible because some of the owners of the railway company were citizens of the city

---

12 The street was rechristened Straße der Pariser Kommune (Street of the Paris Commune) in GDR days, the name it holds today.

and therefore entitled to buy property within the walls. In addition, there was the space; whereas the western section of the city was entirely built up, its eastern half long retained an almost rural character. It did, however, also mean that trains had to drive through a guarded gate upon leaving the station.

During its first years the Frankfurter Bahnhof could therefore hardly be called "urban", as it was situated amidst fields despite being located inside of the city walls. A contemporary drawing shows housing blocks in the background, but these were entirely fictitious at the time; soon, however, reality would more than catch up with the artist, as the Stralauer Viertel, the southwesterly part of Berlin's Friedrichshain district, grew at an rapid pace. Twenty years later, it had exploded into the city's most densely populated neighborhood, packing 330,000 people into an area less than ten square kilometers, and was already gaining infamy by possessing some of Berlin's worst tenement buildings. The east of the city would become known for housing those with the lowest social status; whoever could afford it immediately moved west.

The speed at which the neighborhood around the Frankfurter developed was stimulated by the advent of industry in the eastern part of the city, but also because most people entering the city to settle permanently came from the east: many from Prussia and Silesia but also from the Baltic and from Russia proper. Most of them were poor and few had much in common with the existing population of the city, either religiously or ethnically, particularly Poles and Jews. Many migrants simply stayed where they arrived for the first time around the Frankfurter Bahnhof, which the Berliners soon graced with the nickname "Catholic Station," referring to the religion of the many Poles that arrived through it. Within an astonishingly brief period these groups were to put their stamp on the identity of the city and its population, an influence that can be felt to this day.

Becoming Busy

The station's first year was uneventful, but the situation changed once the line to Frankfurt was extended to Breslau (today Wroclaw) on December 12, 1843. Simultaneously, the company was taken over by the newly formed Niederschlesisch-Märkische Eisenbahngesellschaft (Lower Silesian-Brandenburg Railway Company); not only did this increase the number of passengers but the Frankfurter now also became the unloading point for massive amounts of Silesian coal, needed to heat the rapidly growing city. Unfortunately, the facilities were entirely inadequate to cope with this new stream of goods and the new company kept struggling with the problem until its nationalization, the first of many, in 1852.

This also led to the first phase of an ongoing naming confusion. The new owner started to use the name "Niederschlesisch-Märkischer Bahnhof" (Lower Silesian-Brandenburg Station) along with the official one; after all, a terminus was also a railway company's advertisement for the destinations it served. In spite of nationalization, both names lingered and, as the reputation of the station started to deteriorate, the sharp tongue of the Berliners re-christened it the "Niederträchtig-merkwürdiger Bahnhof" (vile and strange station).

By the 1860s it had become clear that the complex wasn't coping with the immense passenger growth well. Over time it had been updated and even a covered train shed was added but still a lot of the station felt temporary and inadequate; the shed had been improvised, the platforms were narrow and made out of wood and they were far too short, meaning that passengers often had to alight onto the soggy ground.

Frankfurter Chaussee, Station for a Day

Fig. 5.4: King Wilhelm I and Queen Auguste being received by Berlin's mayor Heinrich Philipp Hedemann at Frankfurter Chaussee station (right), 22 October 1861.

Meanwhile, a somewhat surreal episode in eastern Berlin's railway history had taken place in 1861, when the Prussian King (and later German emperor) Wilhelm I returned from his coronation in Königsberg. Since the Frankfurter Bahnhof was situated inside the city walls and Berlin's mayor insisted on leading the new King into the city, a special railway, the Boxhagener Krönungsbahn (Boxhagen Coronation Railway, named after the estate it traversed), was constructed to transport the king to a temporary station near the Frankfurter Tor created especially for the occasion. That building, 12 meters high and containing a 38-meter platform, was only used once.

The city paid for most of the project, including constructing and testing a branch line from Rummelsburg station, but citizens chipped in as well – some literally, by donating sand that could be used for the railway bank, and by allowing the line to use their land. This outpouring of affection demonstrated how the reputation of the previously-nicknamed "Grapeshot Prince" had recovered dramatically since the days of the 1848 revolution, when he had to seek refuge after allowing troops to fire on revolutionaries.

At 11.15 AM on 22 October 1861, the royal train entered the station. After having been welcomed by Mayor Hedemann, the party enjoyed a meal in the similarly impromptu station restaurant before making its triumphant entry into the city through the Frankfurt Gate. A month later, barely a trace of the ad-hoc Frankfurter Chaussee station building was left; the tracks had been removed as well. However, the railway bank was to become the Boxhagener Straße, which years later once again saw rails installed, this time for a tram line. Today, it is being used by Berlin's tram line 21.

A New Station – or Almost

By this time, it had become obvious just how inadequate the Frankfurter Bahnhof's facilities had become. Plans to also use the crumbling complex for traffic from the Eastern Railway to Königsberg and beyond finally set the wheels of state in motion; architect Eduard Römer was engaged to conceive an entirely new station building, while the facilities for handling goods and mail were to be improved as well. A first design was presented in 1865 and quickly green-lit; building started two years later, after the opening of the nearby Küstriner Bahnhof, built for the Ostbahn to Königsberg (see chapter eight), and the temporary replacement for the line to Breslau during construction of the new Frankfurter Bahnhof.

The design deviated from Berlin's default in a number of ways, giving a neogothic, angular twist to the Berlin Rundbogenstil. The arrival and departure sides were of course swapped the right way around, but the *Berliner Breite* for official buildings of 37,66 meters was still respected. The usual office building at the front was omitted and replaced by a transfer table to move the locomotives between tracks, which was hidden behind a garden-like wall and gallery and would give the station a friendly, accessible, yet suitably imposing face.

At least, that is what was *supposed* to have happened. Political meddling meant that compromises had to be made, as plans to raise the tracks a few meters to stop the railway operations from interfering with street-level traffic were deemed too expensive. Moreover, the Prussian minister placed restrictions on both ends of the station; to begin with, the blighted crossing with the Fruchstraße had to remain, placing a fixed limit on the length of station on the eastern side. In itself, this might not have been a huge problem had the minister not also insisted that the old station administration building from 1842 be kept in place. Although this inserted an element of continuity, the decision also set another hard border at the western edge of the plot, further restricting the size of the new complex. Moreover, it nullified much of the intended aesthetic impact of the new complex.

The new building now used up all the space between the administration building and the Fruchtstraße; as a result its 210-meter platforms could only accommodate trains up to 55 axles; however, the very first train which pulled into the new station contained 56. For that reason, the longer trains serving the Ostbahn continued to use a different station, the Küstriner Bahnhof (see chapter eight). The general feeling of dissatisfaction with this solution is demonstrated by the photographs taken of the newly completed station taking conspicuous care to avoid showing the remains of the old one (see Figure 5.6).

Inaugurated on 16 August 1869, the new station nonetheless offered a vastly improved experience to travellers. It was nothing if not opulent, containing spacious rooms for all four classes of passengers and royal waiting rooms both at the entrance and exit of the station – a first for Berlin. It wasn't meant to last. Even more than was the case with the other Berlin termini, the Frankfurter was overtaken by the rapid growth of the city. In 1871, the Ringbahn replaced the Verbindungsbahn, the street-level railway which had connected all the stations for the supply of freight and coal. Built with some foresight, the new "Ring" took a wide arc around the city, and also carried passengers. Although a connection was created to the Frankfurter, there was no real space for the local trains, which meant that an additional platform needed to be improvised.

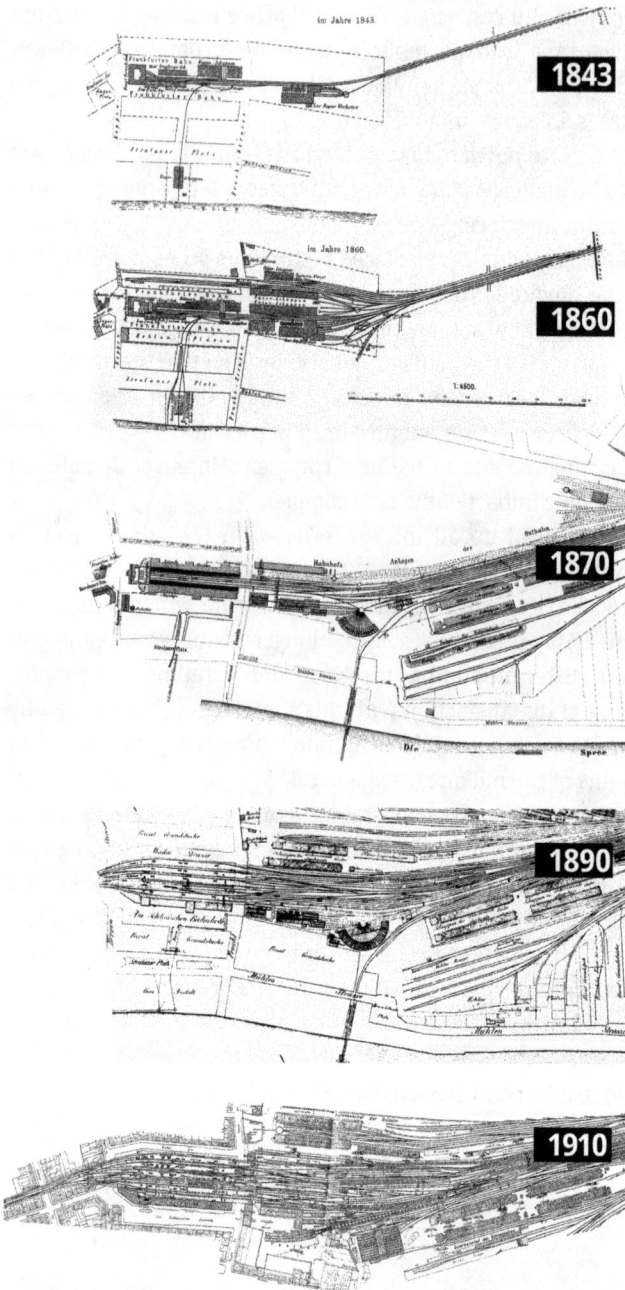

Fig. 5.5: The evolution of the Frankfurter/Schlesischer Bahnhof from 1843 until the beginning of World War One. To scale.

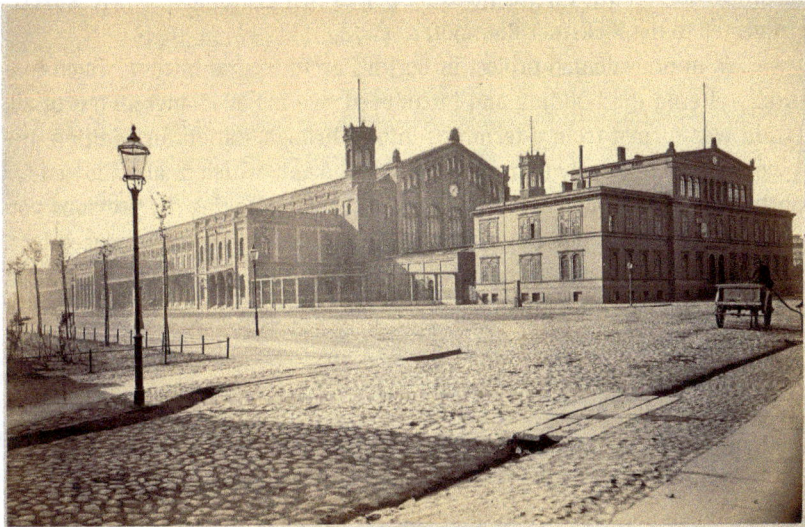

Fig. 5.6: The power of presentation: official photograph taken at the opening of the new Frankfurter Bahnhof (top) vs. the actual situation (bottom).

The "Eastern Station"

By this time, it had become clear that the Ring would not solve all of Berlin's traffic problems. In 1871, not entirely coincidentally the year of German unification, the architect August Orth (the same that designed the nearby Görlitzer and later, new Stettiner Bahnhof) proposed a "local locomotive railway" to run through the city itself instead of around it. This was not the first time a more encompassing and drastic solution had been proposed, but the opposing interests of the individual railway companies had long caused them to resist such plans. The way in which the Franco-Prussian war had proven the military significance of the railway caused a more widespread realization that such infrastructure was too important to leave in the hands of these disparate, private parties; in the long run, this led to rail nationalization, but it also had the immediate effect of removing most public sympathy for the railway companies' complaints. The result, after ten years, was the Berlin Stadtbahn, a railway traversing the city from east to west (see chapter six).

From a fairly early point onwards, the Frankfurter Bahnhof had been suggested as a logical eastern point to connect the Stadtbahn to the other railways headed east. In 1877 a conference was set up to determine how the station – not even eight years old by this time – needed to be adapted to become the "eastern station of the Stadtbahn" and as building began two years later services were, again, diverted to the Küstriner Bahnhof.

It was an unprecedented project in Berlin's architectural history, which fundamentally altered the building and introduced two major changes. First of all, the station was turned from a terminus into a through station. In addition, the tracks were raised by nearly four meters inside the hall, which finally allowed for the construction of a bridge across the Fruchtstraße, removing the previous bottleneck. Much of the old building was gutted, with the saddle roof replaced by a barrel-type one, and the front torn out. The last surviving part of the 1842 station complex, the administration building at the front, was also finally removed to extend the tracks in the direction of the Stadtbahn.

New Wine in Old Bottles

A lot of the old station was left standing due to time and money constraints: the walls and platforms of the old hall and much of the original entrance building with its waiting rooms, counters, and luggage facilities. This way of adapting an old structure wasn't entirely unique in Berlin's railway history, but it was certainly rare; the departures (south) side of the building now became its main en-

Fig. 5.7: The new Stadtbahn hall of the Schlesischer Bahnhof photographed by Hermann Rückwardt, 1882.

trance, with the old arrivals (north) building entirely removed to allow for the construction of a second hall, wider than the old one, used exclusively by the Stadtbahn trains. To emphasize its radically different nature, it also received a new name, and was re-christened the Silesian Station, or Schlesischer Bahnhof.

Passengers could now reach the platforms using street-level tunnels and stairways. Although the Fruchtstraße crossing had ceased to be an issue, the extension of the platforms was done at the western end of the station rather than at the eastern end; long-distance trains were now able to enter the old hall from each side, but only a single track connected the two long-distance sides – so in practical terms it mostly served as a bilateral terminus, and only a single service initially made the full west to east crossing across the city: the London to Moscow express.

Opened by Kaiser Wilhelm I on 7 February 1882, the Stadtbahn proved to be an immediate success, quickly establishing itself as the central traffic artery for the German capital – a status it has never lost. By the turn of the century, the Stadtbahn as a whole easily outdid Berlin's other stations combined in terms of passenger numbers, while the Schlesischer Bahnhof itself was the third most heavily used of all when it came to long-distance travel, after the Stettiner and Anhalter. After 1882, it served not only connections to the east, but also to Hannover in the west, and by extension Belgium, the Netherlands, and France,, which also made the Schlesischer Bahnhof a place where heads of state were received and where the imperial family departed for the western and eastern provinces; consequently, the royal facilities remained as sumptuous as they had been before.

A Busy Place

In terms of surface area, it had now become Berlin's largest train station, even bigger than the mighty Anhalter. However, the removal of the arrivals building to construct the second hall meant that all services now needed to be concentrated in the previous departures wing. A small extension was constructed at the west end of the building but this didn't prevent the station, now the city's second busiest in total number of passengers, from becoming much more crowded and feeling more cramped. Combined with the station's somewhat mongrel appearance, the growing stream of arriving destitute immigrants from the east, and the proletarian nature of the surrounding area, this contributed to its growing reputation as a place where one shouldn't linger, particularly after dark. A sizeable portion of the passenger station was also still used for the handling of mail, a situation that would only be alleviated when separate post facilities were opened in 1908.

Normal passengers had to make do with rather more humble facilities than the royals, and contrary to the Anhalter Bahnhof, whose waiting rooms garnered a notoriety for excellent service and cuisine, the Schlesischer's facilities acquired an altogether more dubious reputation. The third class waiting room in particular became known as a place for conducting criminal transactions, ripping off desperate immigrants and upsetting the forces of the establishment – sometimes by the same people and at the same time. This is where the Ringverein Immertreu operated from, and the surrounding quarter was not much better, earning the station the nickname of the "Drei-Groschen-Bahnhof" (Three-penny station) because that was the going rate for making "acquaintances" in the neighborhood.

Simultaneously, the authorities were also very present at the same station, not only during state occasions but in a more tangible form during World War I, when the Schlesischer served as a command center and all trains running to the eastern front from Berlin and central Germany were routed through the station. That earned it the attention, in the opening months of 1919, of revolutionaries who wished to prevent the military from bringing troops from the east to sup-

Fig. 5.8: The Schlesischer Bahnhof's 3rd Class waiting room in 1930, usually smelling of "unaired people, beer, cheese and garlic. Tobacco ash and street dirt cover the floor" (Julius Berstl, *Berlin Schlesischer Bahnhof*, 1964).

press Karl Liebknecht's communist uprising; about 300 of his Spartacists occupied the building but had to abandon it after the surrounding area fell to government troops (there is some confusion whether the revolutionaries abandoned the station or were forced out by government troops storming it). This wasn't the last time the station would become the focal point of political agitation; its position in the heart of Berlin's "reddest" quarter guaranteed it that.

This wasn't the only battle that took place around the station; in 1928, a kerfuffle took place between members of "Immertreu" and carpenters from Hamburg, who had been working on the construction of the subway (part of today's U5). Over three days, the brawl escalated into an all-out war between the Hamburgers and thugs from various Ringvereine, who considered the "foreign" laborers a threat to their authority. Although almost all of them were acquitted in the subsequent trial, this turned out to be a pyrrhic victory in the long run. The role of the Ringvereine was now public knowledge, and would cause the Nazis to put their leaders into concentration camps shortly after taking over in 1933.

However, neither organized crime nor revolutionary agitation constituted the most nefarious thing that went on at the Schlesischer Bahnhof during the Weimar Republic. That distinction belongs to Carl Großmann, the proprietor of a sausage stand in front of the station; one which seemed to remain suspiciously well-stocked in spite of rationing after the First World War. Großmann, a recurrent sex offender, was arrested on 21 August 1921 when police found the body of a recently murdered woman in his apartment, with the stove in his kitchen containing the remains of at least three other people. Body parts had been fished out of the river Spree near Großmann's house for some time, and it is estimated that his victim count must lie somewhere between 50 and 100 women, mostly prostitutes. Even though the police were never able to prove it, it was widely assumed that the victims of "The Beast of the Silesian Station" must have found their way into his sausages. The third class waiting room's cuisine at the Schlesinger might not have been the greatest, but at least one didn't eat one's own species.

Renovation, Destruction, Reconstruction, and Josef Stalin's Drive-by

By the mid-1920s, the Schlesischer was in dire need of renovation because smoke and soot had eroded the beams supporting the roof of the (newer) northern hall. During those works, the northern wall that had once been part of the old Frankfurter Bahnhof was also removed. The entire southern hall was itself renovated between 1933 and 1938, and by the beginning of the Second World War its roof had been entirely replaced.

Of course, that conflict proved as damaging to the Schlesischer Bahnhof as it was to the other Berlin stations. Uniquely, the building survived the skirmishes of the spring of 1945; while it did suffer extensive damage, particularly the entrance building, unlike the nearby Küstriner Bahnhof it was able to resume service relatively quickly afterwards. By July, the trains could run again.

They had to. Remember that strange impromptu station built for the emperor in 1861? East Berlin's legacy of ad-hoc railway infrastructure to serve autocrats received an even more bizarre episode when Josef Stalin needed to visit the Potsdam Conference in 1945. As the Russian dictator insisted on using his own train and because the Russian gauge is a bit wider than the European one (1520 mm vs. 1435 mm), the entire German track from the (pre-WW1) Russian border to the Schlesischer Bahnhof was re-laid to allow the train to travel to Berlin. Afterwards, it was returned to its old state, at least up to the point where it crossed the new Soviet border.

Intermezzo: A Station on the Side

To the north-east of the station, literally in its majestic brother's shadow, existed a little terminal station that has almost gone missing in history. Today, no trace of it can be found and it has been replaced by a somewhat anonymous collection of buildings. However, for some time, this was where many of East Berlin's workers entered the city. Unlike the Silesian Railway, the commuter line from Wriezen had never been elevated above street level; initially trains had ended at the Küstriner Bahnhof but when that was closed in 1882 the typically ad-hoc decision was taken to simply tack a small, street-level platform onto the big Schlesischer Bahnhof, south of the Fruchtstraße.

Initially simply known as the "Wriezen platform" (Wriezener Bahnsteig), it gained its own little entrance building just before World War One, after a gas explosion had damaged the structure. It was a modest affair, stylistically reminiscent of a scaled-down version of the Potsdamer Ringbahnhof, to the side of the Schlesischer Bahnhof. In 1924, the platform was formally turned into a separate station, and renamed "Wriezen Station."

People alighting from the station had to walk about two hundred meters to enter the bigger station. During the war, the little station was hit just as hard as its neighbor but it was among the first to be used again after the cessation of hostilities, despite the damage; in fact, the only photo we seem to have of the station is in its damaged condition, just after the war. The Wriezen station was patched up when it was over and for decades continued to serve initially as a commuter station and later as a museum for the Reichsbahn, the East German railway company, which exhibited historically significant trains at the station with some regularity.

At this point, there were three stations in Friedrichshain: the Küstriner, already out of action; the Wriezener, still in use but without a real future, and the Schlesischer, also in use and with plans for reconstruction being drawn up. Things were going to get complicated, though. In 1950, the GDR authorities in the Soviet sector of Berlin ordered that names that referred to previous German lands that had become Polish after the war needed to be changed. Therefore, while the Stettiner Bahnhof became Nordbahnhof, the Schlesischer Bahnhof became Ostbahnhof. The problem was, of course, that there already was an Ostbahnhof[13] since that was still the now-ruined Küstriner's Bahnhof's official name.

Simultaneously, the Wriezen station lost its autonomy, as services to and from the town were now handled by Schlesischer/Ostbahnhof. In their wisdom,

---

13 And a Nordbahnhof; see the previous chapter.

Fig. 5.9: Wriezener Bahnhof, somewhat the worse for wear, c. 1946.

the authorities chose to re-assign the name "Wriezener Bahnhof" to the previous Ostbahnhof/Küstriner Bahnhof (Küstrin was now located in Poland, too, so that was out). The passenger station might have disappeared, but there was still a goods station on the site, while the adjacent street, Am Ostbahnhof, was re-christened "Am Wriezener Bahnhof," thus causing people to think that the Wriezener Bahnhof was where it wasn't and creating a quagmire of confusion that exasperated the GDR authorities themselves at times. The fact that the Wriezen Station was situated on a street called An der Ostbahn (and thus sounded very similar to Am Ostbahnhof) only made matters worse.

Steadily crumbling, the station lasted until 2004, when the railway installations to the south of Ostbahnhof were renovated and the remaining platform demolished. The only thing remaining today is the administrative office of one of the two goods sheds, which has been restored to even more than its former splendor; however, unfortunately, the surrounding area lets it down rather badly.

A Main Station for the Capital of the Republic

Unlike most of the other large Berlin termini after the partition of the city, the Schlesischer Bahnhof was still useful since both it and the lines it served were placed inside the Soviet sector.

Fig. 5.10: Berlin Ostbahnhof in 1966.

The station could therefore be reconstructed as the Ostbahnhof (Eastern Station) between 1949 and 1950. The reconstruction was mostly a patch-up job, and the end result still looked much like the old building. The western extension and entrance building were entirely removed, and the main entrance enlarged, but stylistically it still clearly referred back to the building constructed in 1882.

The name "Ostbahnhof" never sat very well with the GDR authorities and in the 1980s it was decided that Berlin's 750-year anniversary provided a good opportunity to turn it into a true *Hauptbahnhof* (Main Station). It had come to serve most of the GDR thanks to the outer ring and by this time it had also become obvious that the 1950 renovation left a lot to be desired; for one thing, it had been a rush job, using low-grade materials that could be pried away from the Soviets but which did not guarantee longevity.

Perhaps worse was that the GDR desired something a bit more representative as its main rail hub, and with fewer links to the problematic German past. This time, the regime was more thorough, removing the entire entrance wing and replacing it with a far larger hall. It also – finally – installed overhead power lines in the station, making it possible for electric trains to enter the train shed on power. Up to that point, they had to coast into the station after losing connection

Fig. 5.11: Berlin (first) Hauptbahnhof during the last days of the GDR, 1987. View from the southeast.

with the overhead electricity electricity just before entering the train shed. Re-naming it Hauptbahnhof (Main Station) was of course also something of a stab at the GDR's West German and West Berlin rivals and a display of confidence by the almost 40-year-old "Farmers' and Workers' state".

Back to Ostbahnhof

As we all know, history turned out differently. A mere three years after East Ber-lin's Hauptbahnhof opened in 1987, the state that rebuilt it had ceased to exist. For some years, it retained the name but, as soon as Berlin started to plan a brand-new main station, yet another name change for the old Schlesischer Bahn-hof became inevitable, and in 1998 it returned to being just Ostbahnhof. A few years later, the entrance hall was again renovated after construction issues from the GDR rebuild were revealed and in 2021 work began on another replacement of the roofs of both train sheds.

The "Ost" now represented the east of the city, however, not so much the con-nections to far-away eastern lands. While this symbolized a real decline in the sta-tion's prestige and status, with its 100,000 daily visitors it can still count itself among the major Berlin transit hubs. However, those numbers are nonetheless far

Fig. 5.12: Surviving windows from Römer's station of 1869 in today's Ostbahnhof.

lower than the more centrally located stations of Hauptbahnhof or Friedrichstraße. As a building, it sits almost forgotten between the commuter hubs of Ostkreuz and Hauptbahnhof; the surrounding area, devastated first by poor construction, then by the horrors of war, and lastly by a singularly loveless reconstruction during the GDR era, offers little or no incentive to visit.

Still, the building offers a unique view into the various phases of Berlin's railway architecture and while a hundred thousand is a lot of people, the large Ostbahnhof can easily handle them, making it a much nicer station to use than the garishly chaotic and always over-crowded Hauptbahnhof. It has survived a monarchy, a republic, and two authoritarian regimes. And of course, with German history you never know: another heyday may yet arrive.

6 The Stadtbahn: Europe's Longest Station

Berlin is like a swamp of air; what it embraces, it will never release. Next to us the Stadtbahn trains glide incessantly, the third classes overcrowded, workers and more workers;
the second class empty; only occasionally we're offered a better image. Below us rattles the
long-distance traffic – the long lines of through trains, the lumbering freight trains. It's closing time. That's when Berlin is at its most oppressive, spitting masses of people. – *Johannes
Richard zur Megede, Von zarter Hand (By a delicate hand, 1899)*

Fig. 6.1: Aerial photograph of the Berlin Stadtbahn seen towards the west, with Friedrichstraße
station visible in the lower middle and the Lehrter Bahnhof seen in the distance, 1933.

On re-opening as the Schlesischer Bahnhof in 1882, that station became the literal
start of the Berlin Stadtbahn, a 12-kilometer stretch consisting of four tracks crossing Berlin east to west. Arguably the German capital's greatest infrastructural
project of the nineteenth century, it soon became Berlin's railway backbone and
has retained that status for the past century and a half. When the old termini became mostly unusable after the Second World War, over time the station came to
replace them for both local and long-distance traffic, with the mere fact that it
could take over that traffic relatively easily is testimony to a rare amount of foresight that led to its construction in the 1870s and 1880s. While none of the stations
along the Stadbahn functioned as a true terminus, any history of Berlin's railways

https://doi.org/10.1515/9783111381879-006

would be incomplete without their inclusion. In a lot of respects, the Stadtbahn functioned as – in the words of Berlin's railway historian Alfred Gottwald – a single station with a 12-kilometer platform; because of that structure the Stadtbahn is an exceptional case, not only for Berlin but for the development of urban railways in general. Furthermore, it is fair to say that the Stadtbahn *made* Berlin.

Before discussing it, we need to get some potential misunderstandings out of the way. Any user of public transport in today's Berlin will be familiar with the term S-Bahn. In addition to the U-Bahn, the trams, and buses, it is one of the mainstays of local transport and the railway service with the greatest coverage in the city. The origin, and even the meaning, of the moniker S-Bahn is debated; some consider it synonymous with *Stadtbahn*, others identify it with *Schnellbahn*, and then there are those who think it really is just a nonsensical term to distinguish it from the *Untergrundbahn* or (more commonly) *U-Bahn*. What makes the situation even more confusing is that while the S-bahn (the service) runs on the Stadtbahn (the railway), it is not the only type of train that does.

The "Verbinder"

The haphazard origins of the first railway connections to Berlin and the practice of building individual terminals for individual lines led to a situation where the city had to be traversed by some other means to travel onward. In an increasingly busy city that was perennially under construction, the logistics of getting people and cargo from one station to the next proved complicated; all goods needed to be offloaded onto carts to be driven through the city center on horse-drawn carts, and then loaded onto cars at another station. It was a laborious process, not made easier when city gates had to be crossed.

By 1848, the number of railway connections to Berlin had grown to five, all bringing people and goods into the city that needed to be transported to other stations. On top of that, Berlin was increasingly becoming an industrial center and the railways were used to bring the city's products to Germany and the world, while the Borsig locomotive factory near the Stettiner Bahnhof had no means to run its products out under (literally) their own steam, which was not only an inconvenience but also somewhat embarrassing. Likewise, the transport of increasingly huge amounts of coal through the city was becoming a logistical problem that was difficult to solve by traditional means.

The Verbindungsbahn provided a solution to these problems, but it was far from perfect. Originally conceived as a horse-drawn connection, it was eventually created as a steam-powered one and partly encircled the city center. Starting at the Stettiner Bahnhof in the north, the single-track railway ran via the Hamburger and Potsdamer stations on to the Anhalter, then (from 1866) the Görlitzer, and, after crossing the river Spree, finally the Frankfurter Bahnhof in the southeast. In particular, those last two connections were important because they gave access to supplies of high-quality coal from Silesia. The plan had originally been to complete the circle at a later stage, but those plans were never realized; for 20 years the *Verbinder* ran back and forth between Berlin's stations, transporting coal and other cargo, while in 1866 and 1870 it also transported soldiers bound for the front in Bohemia and France.

Fig. 6.2: Verbindungsbahn transporting soldiers to the Potsdamer Bahnhof, 1871.

For all its utility, the Verbinder also quickly showed itself to be a bit of a nuisance. Its tracks were all at ground level, meaning that the often-sizeable train of carriages interfered with other traffic on Berlin's increasingly busy streets, while it was also slow, noisy, and dirty. The train largely ran on the inside of the city walls, but the stations it served had almost all been constructed outside of them, meaning that it needed its own, manned gates and often had to noisily wait for them to be opened. In 1864 the Verbinder was therefore restricted from operating during the day, but while this solved the worst of the traffic issues, it also meant that people along the route were kept awake.

Around Instead of Through

The unsatisfying nature of the Verbinder had not escaped the authorities, who set to work on devising a suitable replacement in 1867. By this time, the hated tax wall was being demolished. A cross-town connection was nonetheless never considered; rather, the opposite view was taken, that the *Neue Verbindungsbahn* should circumvent the city entirely, but at some distance (one to six kilometers) and then branch off to the various termini. This way, it was hoped, the inconvenience could be kept at a minimum and also leave some space for urban expansion. The new railway was partially laid out on an embankment of around five meters in height – to avoid the same issues that had plagued the Verbinder but also to give it a firmer foundation on Berlin's marshy soil.

Soon rechristened the *Ringbahn* for obvious reasons, it was opened in late 1871 for goods, and then from 1 January 1872 for passenger services as well. At first, it stretched eastward from Moabit, via Stralau-Rummelsburg (today's Ostkreuz) to the Potsdamer Bahnhof in a semi-circle, with the western section completed in 1877. The old Verbindungsbahn was immediately closed, with the exception of the stretch between the Görlitzer and Frankfurter Bahnhof, needed to supply the Berlin gasworks with coal.

The Ringbahn soon proved to be a huge improvement over the old, rickety Verbinder, not least because regular passengers could now use a transit service to get from one station to the next. It also established itself quickly as a commuting service in its own right, and as such played a big role in the expansion of the city, however, that progress came at a cost: time. In terms of city planning, the idea of circumventing the whole center at a distance might have been a wise decision, but it also meant that to go from one station to another passengers had to travel across a relatively large distance. In some cases, such as from the Lehrter to the Potsdamer Bahnhof, walking could in fact be quicker. Nonetheless, the Ringbahn marked an important stage in Berlin's history, and one that can still be felt today; it helped to develop the city into a true metropolis by drawing it closer to its outskirts and satellite towns, and moreover created a sense of "outside" and "within" that still shapes people's perception of the city today. Its typical, "dog's head" shape turned into an immediately recognizable feature on the Berlin city map.

Across Again

The Ringbahn largely solved the problem of cross-city transportation for goods and made it possible for commuters to easily reach one of the termini around the center of the city. Going into the center itself, however, still required additional transporta-

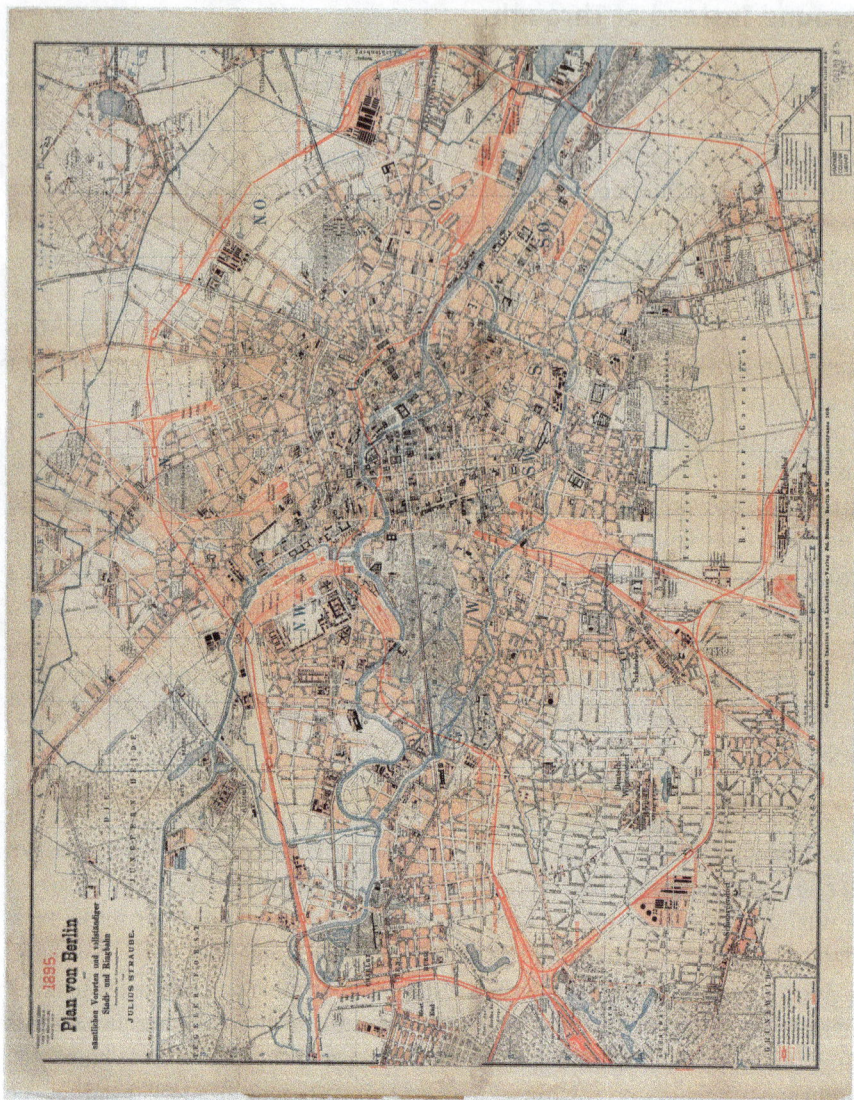

Fig. 6.3: Map of Berlin from 1878 with the recently completed Ringbahn drawn in. Note the still very recognizable "dog's head" shape of the ring on the map.

tion, and while the Ringbahn helped to bring the city together, it did little for the empire's capital as a transport hub within the country; travelers from east to west still needed to cross the dense city center from one terminus to the next. August Orth's solution in 1871 in the form of a "local railway" (*Lokale Lokomotiv-Eisenbahn*) envisioned an elevated connection crossing the city east to west along the river Spree. Not only did long-distance travel play a role in his proposal, but also the alleviation of the city's housing shortage by making its suburbs and satellites more easily accessible. In essence, Orth's concept already corresponded to the Stadtbahn as it would eventually be built, with the plan further elaborated two years later by the engineer Emil Hartwich's proposal for a "Berlin Central Railway"; from this point onwards there appears to have been a general consensus about the direction of the project.

Fig. 6.4: Construction of the Stadtbahn near Alexanderplatz, 1880.

The final decision to build an elevated railway mainly intended for passenger transport across the now imperial capital of Berlin was taken in 1875 and within seven years it would be finished. Although the project had originally been financed by a partnership of state and enterprise, by 1878 the Prussian state had assumed control. As a consequence, some early initiators, such as Orth and Hartwich, were shut out but the project itself moved forward a lot quicker.

One of the fears was that a railway penetrating a densely built-up city center would be prohibitively expensive, but fortunately there was a canal, the Königsgraben, which the city had regarded as something of a headache for a while; having lost its defensive importance some time ago, the foul-smelling stretch of water was regarded as a health hazard in a time when outbreaks of cholera and other contagious diseases were a constant worry. Filling in the canal and building an elevated railway over it solved two problems at a single stroke but, of course, this covered only part of the distance. However, the land between the Frankfurter Bahnhof and the city canal was soon purchased, and the city gave permission to use the northern fringe of the Tiergarten park, which it was in the process of buying from the adjacent city of Charlottenburg. Some commenters doubted whether replacing the canal by a four-track railway meant an improvement for the locals; although smelly, at least the canal was quiet. As part of the city's defense works, it was also not entirely straight; the repercussion of using its trajectory can still be felt in the many bends all trains still need to take along a good portion of its length.

The construction of the Stadtbahn was without question the most important infrastructural project in Berlin during the nineteenth century. It truly transformed the city and the way it was experienced by both its citizens and visitors. Connecting the new line through the city center with the Ringbahn also meant that it could now be divided into northern and southern rings, both of which connected to stations in the city center. For most of its length, the elevated line was built on a total of 731 stone arches, which created a 12-kilometer string of private businesses that contributed to the exploitation of the line. Around the stations, the arches were largely occupied by restaurants and bars. In quieter areas they were used to house other businesses and storage facilities.

Having found a place to build the new line, the obvious decision was to turn the (then) Frankfurter Bahnhof into the easternmost Stadtbahn station. Opened eight years earlier, it needed substantial modification and extension though, since its had previously been a terminus; moreover, a separate hall for the Stadtbahn trains (in the early years called Vorortverkehr, i.e. local trains) was added to create what was now rechristened the Schlesischer Bahnhof, the "eastern station of the Stadtbahn" (see chapter five). The Stadtbahn contained five stations where travelers could take long-distance trains; from east to west, and beginning from the Schlesischer Bahnhof, these were Alexanderplatz, Friedrichstraße, Zoologischer Garten, and finally Charlottenburg. The local trains and long-distance trains drove on separated tracks.

In addition, several stops (Haltestellen) only served local services but were bypassed by the other trains. Here, people could catch connections along the Stadtbahn, to the Ringbahn and so to the other termini in the city. This proved to be an immense improvement in urban transit, since the outskirts of Berlin now gained a

Fig. 6.5: View of the canopy and into the train shed at Alexanderplatz from the east. Photo by Hermann Rückwardt, 1882.

direct fast connection to the city center for the first time. The Lehrter Stadtbahnhof, built atop the railways leaving the Lehrter Station (see chapter 9), was only used for local traffic, so long-distance passengers en route to Hamburg from the east needed to change onto a Stadtbahn train in order to catch a connection to their final destination. While this might have been unpractical, the suffering was somewhat alleviated by the high frequency of the Stadtbahn trains; several east to west lines were combined along the Stadtbahn, leading to as high a frequency as the limits of steam propulsion and security systems would allow.

In terms of usage, the Stadtbahn was a tremendous success. In 1896, just 14 years after opening, its stations served almost 37 million passengers, of whom 31 Million used it for transport within the city, about five million to reach the outer suburbs, and another 1.2 million for long-distance journeys.[14] The easternmost three stations also formed the city's top-three in terms of passenger numbers, although the vast majority consisted of local travelers. It had quickly become the city's prime transport artery.

However, it is important to realize that for long-distance travel from and to the capital, the consequences were not as far-reaching. Some minor schedule changes took place, the most important being that most trains for Hannover and the west of the empire now came to run along the Stadtbahn instead of starting at the Lehrter Bahnhof. The Lehrter became the gateway to the ports of Hamburg and Bremerhaven in the north, while the Hamburger was taken out of service entirely. Apart from trains heading east leaving from Charlottenburg, things stayed more or less as they were. The Stadtbahn was revolutionary for Berlin but mostly because it helped to create a unified transit system for the city itself.

Alexanderplatz

Moving west from Schlesinger Bahnhof, Alexanderplatz was the first long-distance station on the Stadtbahn. Of the four new *fernbahn* stations, three (Alexanderplatz, Friedrichstraße, and Zoologischer Garten) gained a reputation that has become as intricately linked to Berlin's history as it was to their transit function. Interestingly, they represent both subsequent periods in history and in architecture (or rather: architectural renovation) as we move from east to west.

The direction of construction of these stations along the Stadtbahn had been entrusted to the architect Johann Eduard Jacobsthal, who designed Alexanderplatz station himself. The three long-distance stations between the Schlesischer Bahnhof and Charlottenburg shared a roughly similar organization and appearance, with a large hall containing two island platforms. Local Stadtbahn trains used the northern island, with the southernmost being dedicated to long-distance services, with facilities and waiting rooms for local and long-distance services located on the ground floor, beneath the tracks. This was a new setup for Berlin, but the traditional separation of arriving and departing passengers was retained through a somewhat over-complicated system of stairs and tunnels, with facilities for local and long-distance passengers also partially separated. Expecting future expansion of long-distance travel, its passengers were given far more space than the former, which turned out a miscalculation, since they were outnumbered by local passengers four to one. Waiting rooms for the four classes of passengers,

---

14 Source: *Berlin und seine Eisenbahnen*, 1896.

however, were combined, as was the ladies' waiting room. Contrary to the other stations in the capital, the Stadtbahn stations did not possess a separate waiting room for the royal family with the exception of the Schlesischer Bahnhof, which already possessed one.

Fig. 6.6: Alexanderplatz Station, interior. Photo by Hermann Rückwardt, 1882.

The Alexanderplatz and Friedrichstraße stations in particular were similar, although the former distinguished itself by a more monumental entrance on the north side, towards (but not on) the square. Defining Alexanderplatz station's atmosphere was its position in the middle of the eastern city center: near a lively square once named for the Russian tsar Alexander I, next to the infamous "Red Castle," the city's police headquarters, and near to the oldest part of the original city of Berlin, the Molkenmarkt. That area contained palaces of the old nobility but was also (and more significantly) known for its derelict buildings, the worst of which was the so-called Krögel, a series of alleyways of partly medieval buildings with later additions known for their derelict condition and unhygienic circumstances.

Fig. 6.7: North entrance of Alexanderplatz station, c. 1890; view from the northeast.

Today, this part of the city has almost entirely disappeared due to a combination of Nazi construction, wartime bombardments, and GDR city planning, but it was the inspiration for Alfred Döblin's novel *Berlin Alexanderplatz* of 1929. The novel, which is set in the area around the station, became an immediate success upon publication; it tells the story of the worker Franz Biberkopf's failed attempts to make a fresh start after his release from prison and losing his battle against the all-consuming monster that is Berlin and its criminal underclass, ending up in an insane asylum where he finally finds solace. Influenced by among others James Joyce's *Ulysses*, the novel became a staple of Berlin literature, subject of no fewer than three film adaptations, the first of which was immediately banned by the Nazis, along with the book itself.

By the time Biberkopf appeared on the scene, the station had already altered its appearance. There had been criticism from the start, mostly because the train shed was very dark on the inside despite the use of electric lighting. By the mid-1920s, it had also become run down because of its heavy use, combined with a lack of proper maintenance during the war. Luckily, by then Germany's coffers had improved to such an extent that a thorough renovation could be undertaken, in which large modern windows were added to the train shed to improve lighting, and its appearance was further modernized. The entrance and reception areas remained more or less untouched, however.

Fig. 6.8: North entrance of Alexanderplatz station today (Compare fig. 6.7).

The importance of Alexanderplatz had also been increased because the station had become an extensive hub for public transport. In addition to various street-level forms of transport, in 1913 the station was connected with its first U-Bahn line (today's U2). A second line followed in 1927 (today's U8), creating a true transit center for east Berlin, which came at the cost of prolonged building activity around Alexanderplatz, creating even greater chaos than usual for nearly a decade and reinforcing the area's less than stellar reputation.

After bombing caused extensive damage to the neighborhood during the Second World War, it was essential for the Russian occupying authorities and later the East German government to get the station back to working order as soon as they could; because of the city's division, it had now become East Berlin's only sizeable public transport hub. The building was provisionally restored immediately after the war, but received a more extensive renovation between 1962 and 1964 in which the exterior was further modernized; the last ruins of the entrance building were removed and a large glass wall installed, creating the Alexanderplatz station that we know today.

Although some restructuring of the station took place after German reunification, particularly in the inside of the building, it has done little to change its appearance. Berlin still really doesn't know what to do with Alexanderplatz square itself, and several plans have been proposed and then rejected in the past decades. In the meantime, it has developed into arguably Berlin's most important tourism spot, with several malls, close to locations such as the Museum Island and with easy access to the clubs of Friedrichshain. Therefore, Alexanderplatz Station remains an essential feature of Berlin's transport landscape, connecting local and regional rail to East Berlin's extensive tram network above a labyrinthine metro station with three lines that Berliners can't help complaining about. Recently, the U5's extension to Berlin's main station (completed in 2021) has given it even greater importance.

Friedrichstraße: The "Central Station"

Fig. 6.9: Friedrichstraße station in its first iteration, around 1905. Colorized Postcard.

Situated about two kilometers west of Alexanderplatz, the heart of the Stadtbahn as it was opened in 1882 stands the new Friedrichstraße station, located in the city center. Initially titled the *Central-Bahnhof* (Central Station) – the name did not stick – it sat in a truly central location, conveniently close to the junction of the city's main thoroughfare Unter den Linden with the Friedrichstraße itself. The latter was quickly developing into a main shopping avenue, and exuded the proper atmosphere of prestige, which contrasted with the decidedly less upscale vicinity of Alexanderplatz.

The impressive, curved building was conceived by Johannes Vollmer who, like Orth, was also active as a church architect. Its design required more inventiveness than its eastern sister because of its position between the Friedrichstraße and the river Spree, where the platforms stretched out over a bridge and above the water. On the north and south sides, however, it was not nearly as wedged in as Alexanderplatz had been, which allowed for squares on both sides. That also permitted it some spatial breathing space, which combined with the curved shape and the bridge helped to give it a more impressive appearance. The train shed was similar to the one at Alexanderplatz but, because of its broad, curved form and the lack of directly adjacent buildings, the windows let in much more light. In addition, the image of steam trains crossing the busy street to enter or leave the station above the busy city traffic soon became something of an icon of the city and its bustling traffic. These reasons all helped to turn Friedrichstraße into arguably Berlin's most beloved station.

Fig. 6.10: The image of trains on the bridge leading into Friedrichstraße station above the hectic Berlin street traffic quickly developed into an iconic image.

What helped was that although it was even busier than Alexanderplatz in its total number of users, they were divided up much more evenly between commuters and long-distance passengers. For that reason, it felt much less busy and cramped for the former, while still remaining spacious enough for the latter. Despite being slightly shorter than Alexanderplatz, it felt bigger. Nonetheless, the same problems of over-use and subsequent wear and tear reared their heads.

Already before the First World War, demolition work on the old train shed had started to create a new one. Due to wartime conditions, actual construction could not begin until 1919, and then lasted for a full six years. In the meantime, the plans for the renovation had been adapted, resulting in a much more modern design than had originally been intended. During the renovation, the entire quay-side construction was overhauled and strenghened, and a third platform added for Stadtbahn traffic on the north side, with a separate roof. Although its renewal was completed earlier than that of Alexanderplatz, it was much more thorough and perhaps therefore ended up looking more modern.

Like Alexanderplatz, Friedrichstraße developed into an important hub, particularly for commuters. The U-Bahn had reached the station in 1923 in the shape of the C line (today's U6 and part of U7) but even more importantly, the opening of an underground S-Bahn tunnel in 1936 turned the station into the main rail hub for the city center. Passengers could now make use of a fast connection from the center to other termini and the suburbs in the north and south, instead of relying on the Ringbahn or slower forms of transport.

Berlin bei Nacht:
Bahnhof Friedrichstr. und Central-Hotel.

Fig. 6.11: Cabs waiting in front of the renovated Friedrichstraße station, 1926.

The situation at Friedrichstraße station changed dramatically with the construction of the Berlin Wall in the summer of 1961. Its position on the border between East and West Berlin, and its combination of S-Bahn, U-Bahn, and rail traffic, put the station in a unique position. The GDR authorities had devised several transit points for street traffic, the most famous being Checkpoint Charlie further south on the Friedrichstraße. Those choosing public transport to cross the zone border had to pass through Friedrichstraße station and the border control building next to it, with this structure becoming popularly known as the *Tränenpalast* or Palace of Tears because of the many dramatic goodbyes.

As far as train traffic was concerned, Friedrichstraße now became a terminus, but a two-sided one. On one platform, trains from the west (the so-called Inter-zone and Transit trains) would end; on the other, S-Bahns from both East and West Berlin terminated.[15] Access to the international platform was obviously strictly guarded. Below ground, the same happened to the U-Bahn and S-Bahn connections. Since the old center of Berlin, now in the east, jutted out into West Berlin like a peninsula, two U-Bahn lines needed to cross beneath it to get from one portion of the west to another. For one U-Bahn line (today's U6) crossing the city center from north to south, Friedrichstraße was the only stop in East Berlin; the trains would not halt at the other, now disused stops (the so-called "Ghost Stations"), which were patrolled by East German security personnel. This experience, along with the Wall's forbidding reputation, granted the station a distinctly sinister image.

After German re-unification, the border station was obviously closed. Friedrichstraße station itself received a make-over, which restored it to its 1925 condition, on the outside, at least. The inside of the building was gutted and renovated in a rather loveless fashion. Today, it is still a very busy station, and made to look more busy by its compact nature (unlike the labyrinth beneath Alexanderplatz). However, it is not quite as important as it once was and plays second fiddle to Alexanderplatz as the center's traffic hub. Post-Wall developments such as the failure of the Friedrichstraße itself to develop into Berlin's main shopping avenue, and the opening of the nearby U5 underground station, have taken away some of its importance for local traffic. Nonetheless, it is arguably Berlin's most beautiful station building today, at least on the outside, and a good example of once-ubiquitous tile-decorated station façades.

Zoo Station and its Children

Proceeding further west along the Stadtbahn to Zoologischer Garten Station, passengers arrived in "Germany's richest city," Charlottenburg; a separate city, it had grown almost as rapidly as Berlin itself, and by the 1880s its built-up area bordered

---

15 This was the usual practice; however, some trains bound for Copenhagen, Warsaw and Moscow might continue after a thorough inspection.

that of the capital. Bahnhof Zoo, as it was soon informally named by the public, was built close to the Tauentzienstraße and Kurfürstendamm, wide avenues that contained shops, cafés, and restaurants at least as stylish as any on the Friedrichstraße or Unter den Linden.

Although Bahnhof Zoo was similar to the stations at Alexanderplatz and Friedrichstraße in many respects, there were also some significant differences. The overall design of the station, while stylistically related, was not quite as grandiose, with a shorter train shed and a separate, lower roof over each platform. Moreover, long-distance and local passengers were separated more strictly than on the other Stadtbahn stations, mostly because of the volume of expected daytrip traffic to the adjacent Berlin Zoo, which the station was named after. For that reason, long-distance travellers used an entrance at the southern end, whereas local passengers could leave and enter opposite the zoo entrance.

Fig. 6.12: The original train shed at Zoologischer Garten, 1915.

In 1902, Zoologischer Garten was the first railway station to be hooked up to the Untergrundbahn (or U-Bahn), the subway connection crossing the city east to west.[16] This fast new cross-city connection further intensified traffic, the more so since it connected directly to the Potsdamer Bahnhof complex and made it an obvious connection for transfers between the two stations. It also became apparent that the rapid development of the "New West," as Charlottenburg came to be known, demanded that it receive a proper facility for long-distance travel as well. Plans took a form similar to those implemented at Friedrichstraße, with the construction of an additional platform and train shed for local traffic so that the other platforms could then all be used for long-distance connections.

The First World War and subsequent economic and political troubles got in the way of executing these plans for a while, but the upcoming Olympic Games of 1936 provided a good excuse to get started with the renovation, as Zoo would be a logical transfer point to the Olympic park in the outer west of the city by both S-Bahn and U-Bahn lines. From 1934, the station was mostly torn down and rebuilt, resulting in a much larger buidling that looked even more modern than Friedrichstraße station and was dominated by steel and large glass surfaces. The new S-Bahn hall was built partially adjacent to it, with most of it on a road bridge over the Hardenbergstraße, further to the southwest. As a consequence, only the southern entrance vestibule offered access to both parts of the station; the northern entrance, although originally the S-Bahn entrance, now only served the long-distance station. That skewed structure made, and makes, the station a somewhat confusing place to navigate in spite of its small size.

Bahnhof Zoo survived the bombs of the Second World War with relatively little damage, contrary to most buildings around it. It was to achieve its most lasting fame after the War, and just like Friedrichstraße mostly as a consequence of its location. To begin with, it turned into the *de facto* main station for the western sectors of the divided city, the main place of arrival for the inter-zone trains from West Germany. After the Berlin Wall had been erected, it became one of only a few entrance and exit points for the isolated city. Unfortunately, two million people using a smallish station originally constructed as a suburban stop took its toll on the building, with both the station and the area around it quickly gaining a reputation as a rather rundown part of town, the complete opposite of the opulent quarter it had been at the time of the station's opening in 1882.

Few contributed more to that notoriety than Christiane Felscherinow, who has become most famous as Christiane F. In a series of interviews, she told jour-

---

16 For the most part, this first U-Bahn connection was an elevated connection (Hochbahn). However, the city of Charlottenburg forced the company to go underground for the stretch crossing its territory.

Fig. 6.13: Bahnhof Zoo in Christiane F's time, 1979.

nalists about her life as a teenage, heroin-dependent prostitute at the station, with a subsequent book published under her name (although ghostwritten) in 1978 as *Wir Kinder vom Bahnhof Zoo (We, the Children of Zoo Station)* which became an instant commercial and critical success. Translated into English as *Christiane F.: Autobiography of a Girl of the Streets and Heroin Addict*, its central theme of an individual that is chewed up by an urban dystopia is eerily similar to the fate of *Berlin, Alexanderplatz's* Franz Biberkopf. In 1981, the book was adapted into an uncompromisingly bleak film, which helped to turn the station into something of a tourist attraction for other teenagers, further aggravating its drugs and crime problems to the alarm of Berlin's health authorities. This solidified a reputation which "Zoo" couldn't easily shake off.

The situation changed somewhat after the fall of the Wall. The song *Zoo Station* by Irish band U2 (sharing a name with, although not named after, the subway line that runs below the station) became a minor hit, referring to the history of the newly united city and the area rather than its murky drug-related past. People visiting the station for a Christiane F.-like experience today will likely leave disappointed; robbed of its unique Cold War status, Zoologischer Garten has since sunk back into suburban insignificance after the opening of Berlin's new Hauptbahnhof in 2006. It is a much cleaner place than it was, and still serves regional and local train traffic, but being so close to the main station means that tourists mostly converge elsewhere.

Fig. 6.14: Bahnhof Zoo at today. The long-distance hall (Fernbahnhof) is to the right, the S-Bahn hall on the left above the road.

Charlottenburg

Charlottenburg Station marked the western end of the Stadbahn, but it was by no means the end of the line. Here, like at Stralau-Rummelsburg in the east, the Stadtbahn connected to the western Ringbahn but also to the railway lines towards Magdeburg and Belzig. The station built for the Stadtbahn on the north side of the tracks was quite different from the other stations along the line; positioned in the still largely undeveloped far west of the city of Charlottenburg, it mirrored the rural feel of the area, in marked contrast to the more urban image of the other stations. Rather than the intricate two-story structures elsewhere, this was a far simpler affair, with its four platforms built on an embankment rather than on stone arches.

Stylistically, it somewhat resembled the temporary, *faux*-half-timbered structures that were also used at the Stettiner and Anhalter Bahnhof during their renovation, sharing their Swiss chalet-like character despite being constructed out of brick. Space wasn't quite as restricted as it had been inside Berlin's built-up area, which allowed for a greater number of platforms where services could already be split up for their various destinations; therefore, it was the largest of the Stadtbahn stations in terms of capacity. However, because of its architecture it retained something of a provisional character.

Fig. 6.15: Charlottenburg Station in the 1920s.

Fig. 6.16: Charlottenburg Station today.

Tragically, that turned out to be the right impression since it was the only Stadtbahn station not to survive the Second World War. Heavily damaged by bombs, a patch-up job had to make do until a new building could be opened in 1971. Because long-distance trains no longer stopped at Charlottenburg, one platform of the station could be reserved solely as a terminus for the British allied forces in Berlin as an exterritorial zone. In West Berlin, the Americans possessed such facilities in Lichterfelde-West, and the French in Tegel. After German reunification regular train services resumed, while the British platform returned to German sovereignty. The S-Bahn station was entirely reconstructed between 2003 and 2006 and moved 200 meters to the west, now aligning with the platforms for the local trains and making it easier to change to the U-Bahn.

That British connection is perhaps also the most interesting way that the station's history is tied up with that of the city. Never truly losing its suburban character, Charlottenburg failed to elicit the same cultural response as the other Stadtbahn stations; if it is to symbolize anything, it is perhaps the somewhat uninspired reconstruction of large parts of West Berlin during the Cold War. Today, it

Fig. 6.17: Karl Wendel, Am Bahnhof Zoologischer Garten (1915).

serves the S-Bahn and regional trains, but it is one of the less important Stadt-bahn stations, and the one where its historical origins are the least tangible.

In a city where much of the development of its transport systems took place in a spirit of exasperation, the story of the Stadtbahn is one of true success. Not only did it help to craft a true metropolis out of the urban upstart that was Berlin but it also turned it into a different city in terms of its appearance and experi-ence, for which it was rewarded in literature, in art, and in song. Few are contest-ing that today it sometimes shows cracks at the seams as train traffic grows ever more dense, but it is still impossible to imagine the transport landscape, or for that matter any landscape, in Berlin without it.

7 Hamburger Bahnhof: More than the Sum of its Parts

Fig. 7.1: Hamburger Bahnhof today. The circular greenspace was once occupied by the locomotive turntable.

With the Hamburger Bahnhof, we return to the 1840s, and the oldest still extant remains of station architecture in Berlin. That the building itself is still among us is something of a miracle, because it really should have disappeared almost a century and a half ago. But we are fortunate that it is still amongst us, because the opening of the Hamburger Bahnhof in 1847 marks a crucial point in the development of railway stations in Berlin, and the creation of a specific identity for this type of building.

Although it is frequently mentioned as being the oldest remaining station building in Berlin, that is only partly the case. That should be understood literally: only a section of the buildings we see today was ever a component of the original station. The train shed, as well as both wings to the side of the main structure, were built much later, while a substantial part of the original complex was torn

https://doi.org/10.1515/9783111381879-007

down. However, the distinctive middle section with its two towers has been part of the Berlin landscape since the station serving Hamburg first opened.

The First Gateway to the North

By 1845, Berlin possessed railway connections to the south and southwest through the Potsdamer and Anhalter stations, to the east via the Frankfurter, and to the northeast via the Stettiner. What it lacked was a connection to the Free City of Hamburg, the German states' third-largest city and also its second-largest port at the time (after Bremerhaven). First attempts to organize a railway between Hamburg and Berlin had preceded the opening of Prussia's first railway in 1838. The reason it took so long to establish this crucial transport link had much to do with the political constellation at a time when Germany was still politically fragmented; the new railway company had to contend with authorities not only in Prussia but also in Mecklenburg, Schleswig-Holstein, the independent city of Hamburg, and even the Kingdom of Denmark, each of which took their sweet time in anticipation of the others. An agreement could finally be reached in 1841 and three years later, construction on the line began.

The line to Hamburg covered some 300 kilometers. Since it traversed a mostly empty stretch of land, the number of stops remained limited and the trajectory of the line could be kept largely straight; both these circumstances allowed for a relatively fast journey, and there was a good reason why this line was chosen to set a series of speed records almost a hundred years later. Still, in those early years the journey could last as much as nine hours.[17]

After some argument over the location of its Berlin terminus, the Berlin-Hamburg Railway Company opted for a spot in front of the Hamburg gate on the Invalidenstraße, one of northern Berlin's main thoroughfares but at the time still relatively rural. Although this made it easy to reach, it also caused problems since the area was still largely a swamp, and providing a solid foundation for the buildings and the tracks would turn the project into an expensive one.

---

17 Today, it's just under two hours when taking a high-speed train, but journeys can still take over four hours on slower connections.

The Ideal Railway Station

The design of the entire project was put in the hands of *Baumeister* Friedrich Neuhaus. In addition to overseeing the construction process of the line, Neuhaus personally designed each of the 24 stations on the line. As a consequence, and contrary to the station buildings along previous railway lines to Berlin, this gave them a fairly uniform appearance. It also saved cost, since the same elements could be re-applied elsewhere. Neuhaus' template was still a neo-classical, Schinkel-esque one similar to the one used in the first Potsdamer and Anhalter stations, using repeating, square forms.

At the Hamburg end, an existing station was adapted to also serve the new line, but the Berlin terminus was entirely new. Of course, these termini were conceived grander than the intermediate stations on the line. Each possessed a façade containing two large doors flanked by a set of large towers. The basic concept of the Berlin went back to 1838, the year Berlin's first station opened. The *Architectonisches Album* of that year contained an article, by the architects Stüler and Strack, discussing the ideal requirements of a new railway station for the Russian capital, St. Petersburg. The authors began by acknowledging that there were not yet any standards for these newfangled station building things, before laying out some of their own: a covered entrance hall, allowing passengers to board trains in comfort; opposite, a similar hall for alighting, with a square in front of it for coaches to transport the passengers into town. The tracks were to be placed between these halls; a portal building with towers, symbolizing a city gate and containing various "necessary spaces", including offices and living quarters for employees of the railway company inside the station building; and depots, workshops, and storage sections in various supplementary structures placed away from the passenger terminus.

The end result already looks suspiciously like the Hamburger Bahnhof did when it was finished nine years later, including the towers with a connecting gallery and the locomotive turntable. The innovations relative to Berlin's older stations were obvious; travelers were offered significantly more comfort, not only when entering trains but also when waiting for their arrival, while those arriving in the city could more easily find further transport.

There were conspicuous differences as well. Neuhaus basically turned Stüler and Strack's floorplan upside down, with the largest administrative building at the front instead of at the rear. Two gates at the front allowed the engine to leave the structure rather than stay inside, thereby preventing its smoke and soot from engulfing and poisoning the passengers. The engine, when detached, could be turned on a turntable at the front of the station, from where it could be driven away through the other gate. Although this arrangement sounded better than it worked in practice, it nonetheless shows the degree to which thought was put into the station's design.

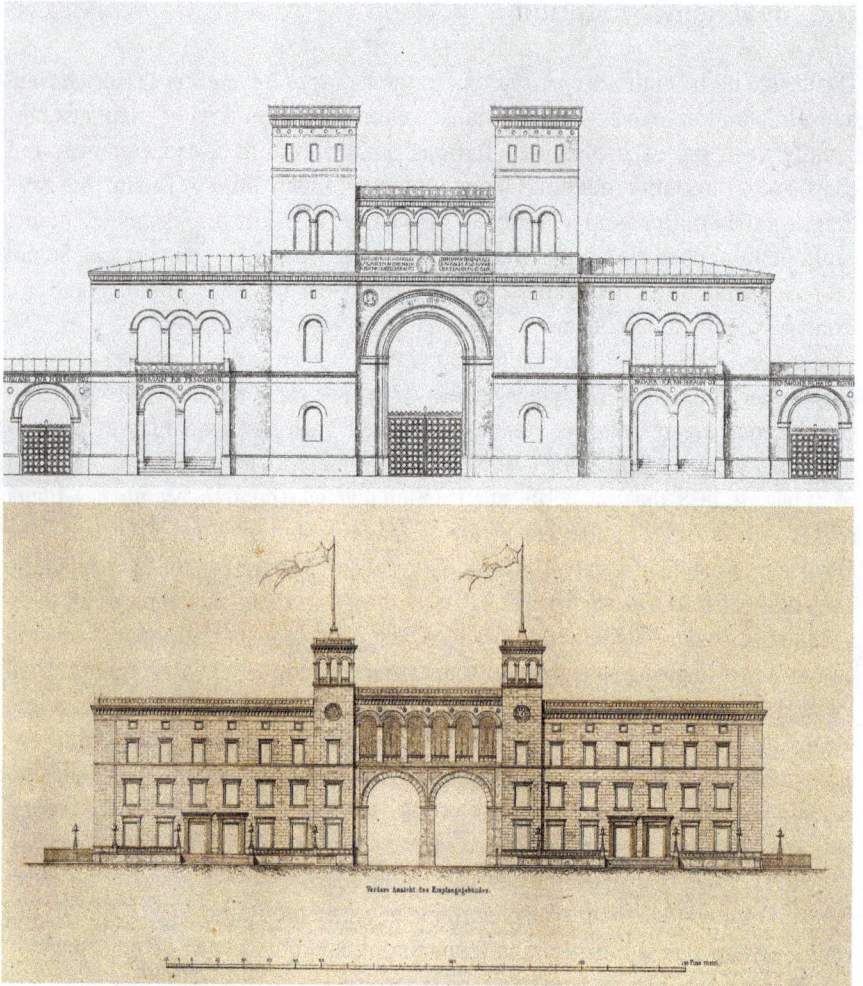

Fig. 7.2: Stüler & Strack's ideal station façade, 1838 (top) and Neuhaus's design for the Hamburger Bahnhof, 1845 (bottom).

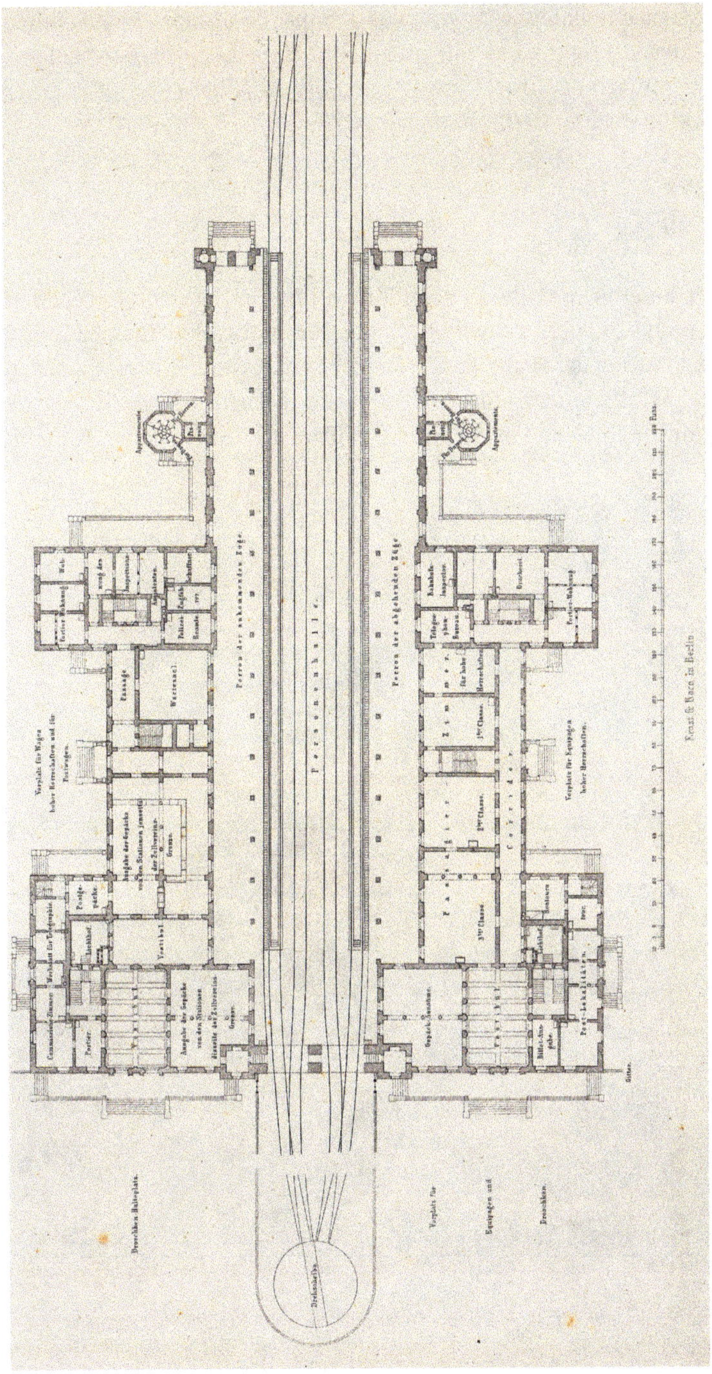

Fig. 7.3: The layout of the Hamburger Bahnhof.

The exit and entrance building were swapped around, because in the meantime Prussia had settled on right-hand rail traffic. But broadly speaking, Neuhaus adopted most of Stüler and Strack's design and organization cues – the latter would in turn influence later Berlin stations as well.

Transport Hub

The Hamburger Bahnhof was the first building in Berlin that truly resembled what we think of as a modern railway station, offering travelers the possibility to wait for their journey or alight under the cover of a roof. This was a sharp departure from the "stationary umbrella" of earlier designs. It was also a very large building for the time, and by far the city's biggest railway station, contain-

Fig. 7.4: The Hamburger in service, c. 1865. Somewhat speculative restoration of the original color scheme on the building.

ing a large train shed (100 meters long, 18 meters wide, and 13 meters high) forged out of cast-iron beams, generous waiting facilities for the three classes of passengers (not counting the royals) and squares on either side for coaches to deliver and collect the passengers. The auxiliary buildings, which included decent goods facilities, were located some distance away from the main terminus.

As such, the new station was the first thorough attempt at an integrated "transport hub" for the Prussian capital. It possessed dedicated sections of the building for entrance and exit, facilities for local transport (mostly coaches and cabs), a luggage facility and comfortable waiting rooms for the three classes of passengers. No longer were passengers forced to face snowstorms or downpours immediately after alighting, or go in a desperate search for further transportation.

Fig. 7.5: The Hamburger Bahnhof as seen from the north, c. 1865. Note the two open gates at the front of the building. The branch to the right connects to the Berliner Verbindungsbahn (Berlin Connector Railway).

Unfortunately, they were still subjected to gusts of wind. For all its cleverness, the Hamburger did come with some drawbacks too, showing just how experimental these structures still were. First of all, those two gates made the hall awfully drafty. Minor adaptations were implemented over the next years to cope with both the growth in passenger numbers and rolling stock size. The turntable was removed in 1874 and replaced by a sliding bed in front of the building. This allowed both portals to be covered with glass, markedly improving the climate inside. An obvious downside was that the smoke could no longer easily escape the train shed. The separation between arriving and departing passengers was still as strict as ever, with no common platform. In practice, however, this hardly ever became a major issue because the station only served a single destination.

The neighborhood around the new station was slow to develop its own identity; in fact, many will claim it still has not really assumed one today. Standing in-between several industrial installations, an army barracks complex, a prison and a hospital, it could be termed a liminal space, existing between other, more clearly defined parts of the city without belonging to either. As Berlin grew in size and density, a working quarter developed to the building's north, and several government and university buildings were constructed to the east, but the station itself continued to exist inbetween these, and largely on its own. The Lehrter Bahnhof, built nearby twenty years later, would also suffer from this problem.

Redevelopment

Like almost every public structure in Berlin, the new station was soon over-whelmed by the increase in passenger turnover as a result of the city's explosive growth. As modern as the Hamburger Bahnhof might have looked in 1848, the speed of developments was such that, during the planning of the next generation of Berlin's stations fifteen years later, it already looked hopelessly outdated. Furthermore, when the line to Lehrte and Hannover was hooked up to Berlin's rail network in 1869, the Berlin to Lehrte railway company opted to build an entirely new station for connections to the north, rather than rebuilding the Hamburger jointly with the Hamburg company. For the time being, the Hamburger therefore remained in service, receiving a renovation in 1877 that lengthened the platforms to 140 meters so it could be used for 12-axle trains. This was left to Neuhaus' son Max, who took care to respect his father's design for the original building.

However, after the construction of the Stadtbahn in 1882 most traffic to Lehrte and on to Hannover was handled by the new line from the Schlesischer, which freed up the Lehrter Bahnhof as the station for Hamburg. The 1884 nationalization of the railway line was the nail in the coffin for the Hamburger; sched-ules were rationalized and simplified, and the station found itself out of a job.

But unlike the Küstriner Bahnhof, where no one really knew what to do with a huge abandoned building, the Hamburger continued to do what it had done since the beginning: provide living quarters for railway (now state railway) per-sonnel. The area was rife with railway installations, and to have staff housed in the vicinity was still tremendously useful.

For about two decades, the station existed as an apartment building. The train shed platforms, and tracks were removed, their space now occupied by an (apparently quite pleasant) common garden. Unfortunately for them, this ideal situation would not last.

The First Museum

Plans to found a Prussian transport museum had been floating around since the 1870s. At one point, it was even considered using the then brand-new Lehrter station as an exhibition space, since it was already situated next to an exhibition area. The question came to a head when no new home could be found for the many expen-sively made exhibits Germany had sent to the 1904 Louisiana Purchase Exhibition in St. Louis. Within months of the issue becoming a political one, the old Hamburger station was chosen as the venue for the capital's Traffic and Construction Museum (Verkehrs- und Baumuseum). Plans drawn up, and the tenants evicted.

It is important to realize how much pride the German and Prussian authorities took in the presentation of German science and engineering in the early twentieth century. During these years, the empire undertook the creation of the German MU-seum of Masterworks of Science and Technology (Deutsches Museum von Meister-werken der Naturwissenschaft und Technik) in Munich, which opened in 1907. In Dresden, a national Hygiene Museum was in the works, and the first initiatives to-wards the foundation of the national society for the promotion of scientific re-search (the Kaiser Wilhelm Society) were taken. Surely no effort was to be spared in erecting a museum dedicated to transport, one of Germany's proudest technical areas of achievement?

As the old train shed had been removed during the station's conversion to an apartment block, a new one, somewhat lower than the original, was installed as the main exhibition space.[18] Further exhibits were set up in what used to be the

---

18 This new shed is often assumed to be the original one, further adding to the lore of "Berlin's oldest preserved railway station."

Fig. 7.6: Kaiser Wilhelm II being received for the opening festivities at the Hamburger Bahnhof, 14 December 1906.

entrance and exit buildings, with the museum not only dedicated to transport technology but also to waterways and construction, which were exhibited in the rooms next to the shed. The portals at the front now became the museum entrance (as they still are today).

On December 14, 1906, the new museum was ceremonially opened by Emperor Wilhelm II. From the outset, it proved to be a huge success and a massive audience magnet, however, soon a very Berlin-ish problem again reared its head. As new exhibits continued flooding in, a lack of space became an issue, prompting the construction of two wings (in 1910 and 1915–16, respectively) at the front of the complex. Unlike in some other cases the architect of these new extensions took great care to have them blend in with the older building. Today, it is difficult to make out that they are over 50 years younger.

Over the next 30 years, the collections kept expanding, and the museum became one of Berlin's most beloved venues. During the Nazi years, another renovation was undertaken, but they might as well not have bothered. With the hostilities of World War came the bombs; situated in the middle of a rail yard and assorted industry, the museum's fate was sealed.

Starting in 1943, the old station was hit repeatedly. The shed (and the objects kept inside) miraculously survived, but the original exit and entrance buildings on the side suffered; the entire eastern flank of the building had burnt out to a

Fig. 7.7: New mock train shed of the Transport and Building museum, 1906. View towards the north; what looks like the back wall was (and is) in fact a large glass window.

shell by the time the Battle for Berlin started. At the end of April 1945, a confrontation between Russian and German troops ensued over the bridge adjacent to the building; because it stood somewhat away from the road it was spared wholesale annihilation, but took multiple hits nonetheless.

To make matters worse, many of the smaller objects were plundered from the ruin after the war – by Russian soldiers and Berliners alike, it seems. Scarcity had turned the metals of which many objects were made into sought-after commodities, but wagons were also stripped of fabrics and even furniture to redecorate the many ravaged houses in the city. Many of the larger objects were shipped off to the Soviet Union, including an imperial wagon that remains missing until this day.

From Cold War Madness to a New Future

What followed next is a cornucopia of absurdity typical for Cold War Berlin. After the war, direction of the German railways was granted to the Soviets and, therefore, to the German Democratic Republic after it was founded in 1949. The now-GDR Reichsbahn continued to own the museum, since it was formally still classi-

Fig. 7.8: The Hamburger Bahnhof after the Battle for Berlin, 1946. In front is the monument for railway men fallen in the First World War.

fied as a station building. However, the building itself was situated in the British sector of West Berlin.

When the West Berlin authorities took possession of their sector's railway infrastructure in 1953, everyone forgot about the Hamburger Bahnhof. As a consequence, what remained of the building was left to decay, with only British military personnel allowed to enter the museum. The structure itself continued to deteriorate into the 1980. However, there was an important upshot; due to this situation the building was saved from West Berlin's post-war demolition drive. In 1984, control over the S-Bahn was granted to the western authorities, who now also came to own the museum for the sum of three million Deutschmarks, a welcome injection of funds for the East German authorities.

This allowed for some basic maintenance work, but by that time the Hamburger's time as Berlin's traffic museum had already ended. In 1982, a new German Museum of Technology was opened to house many of its collections. Rather than using the crumbling ruin of the Hamburger, the new museum utilized the old

Fig. 7.9: The display shed of the Hamburger Bahnhof today; view towards the south.

roundhouses of the Anhalter Bahnhof locomotive depot to display its railway collection; with that, the Hamburger ceased to have any real purpose.

Just as had been the case with the transport museum, it was a collection waiting for a museum that led to the Hamburger becoming re-purposed, but this time as an art museum. In the 1980s, the art collector Erich Marx had donated his collection to the city of Berlin. Looking for a suitable venue, the Berlin Senate decided to house it in the abandoned station.

While renovations were going on, the situation changed again in the early 1990s, after the fall of the Berlin Wall. From its peripheral position in a little-loved corner of an isolated island city, the Hamburger suddenly became a prominent building in the frenetically re-developed new center of the German capital.

This opened up new possibilities. When German President Richard von Weizsäcker opened the new museum in 1996, it had been transformed. Architect Joseph Paul Kleihues, like his predecessors, had the good sense to remain faithful to Neuhaus's original style. Inside, however, it looked little like the original building, with the exception of the eastern wing, which still breathed – and still breathes – some of its Wilhelmine spirit.

What you see today is a historical patchwork, and in spite of appearances little of the original station remains. Yet as an ensemble it works very well, because each of the parties that added, extended, amended, and renovated took care to respect Neuhaus' core; a respect which, certainly in Berlin, is anything but obvious. Still, a lot, including the original arrival and departure halls, has been lost.

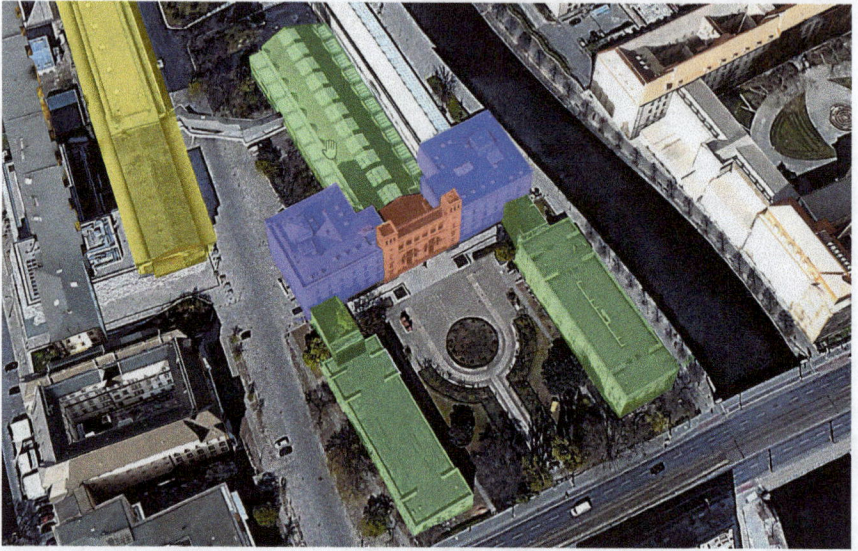

Fig. 7.10: The building phases of the current Hamburger Bahnhof. Red: original building. Green: added in phases as part of the museum conversion (1904–1915). Blue: original building, but destroyed in World War 2 and reconstructed between 1992 and 1996 as part of the second conversion. Yellow: former goods facilities ("Rieck-Hallen").

Unfortunately, the status of the station building remains uncertain. Although the SPK (Stiftung Preussisches Kulturbesitz) exploits the museum, little appears to have been officially put to paper. It has no official lease, it seems, nor does it pay any rent. The owner of the complex, an Austrian investment firm, is currently building new housing on the site of the old Rieck halls, and although the state is in negotiations, at the time of writing nothing has been concluded yet.

8 The Görlitzer and Küstriner Stations: A New Generation

Fig. 8.1: Ostbahnhof (Küstriner Bahnhof) shortly after completion, 1868.

The stations we have discussed so far mostly had their roots in the first rush of railway development that took place in the decade after 1838. During the 1860s, however, a new wave of took shape in Berlin, made necessary by the kingdom's development and thereafter facilitated by Prussia's victories in the three wars of German unification. Largely paid for by the victories over the Danes, Austrians, and French, and necessary both for military and industrial purposes, they sought to tie new Prussian lands to the capital.

This manifested itself in the redevelopment – or the intention to do so – of existing structures, but also in the construction of entirely new infrastructure. During these years, three entirely new stations were erected: the Görlitzer Bahnhof (1866), the "old" Ostbahnhof or Küstriner Bahnhof (1867), and the Lehrter

https://doi.org/10.1515/9783111381879-008

Bahnhof (1869). Of these, the Lehrter was without question the most important, not least because today Berlin Hauptbahhof sits in its location.

In a way they created a crucial new direction for the development of railway architecture in Berlin, mostly because of the lukewarm reaction to the design of the first two and the much keener reception of the architecture of the third. One aspect of these stations drew widespread criticism, and none more vocal than from architects; like most monumental nineteenth-century buildings, the Berlin termini were all styled using designs based on historical examples. The favored template that the Prussian state (and many private parties) drew on was that of the Italian renaissance Palazzo. As a consequence, buildings tended to look quite similar irrespective of function.

In the case of a station, the building also posed a stylistic contrast to the train shed behind. While this might have been acceptable two decades earlier, in the meantime architects had made steps to integrate a building's function into its design. The railways were a source of pride, moreover, and not something to be hidden from view: station buildings were beginning to be used to add grandeur to the city. The way Baron Haussmann used the big Parisian termini to bookend his new urban avenues was becoming noticed in Berlin, too.

Architects, who described the Görlitzer and Küstriner's neo-renaissance buildings in front of the train shed building as a "cake" complained that it reflected nothing of their structure or function. Behind the façade, the setup was fundamentally the same as that of earlier stations, with separated services for arriving and departing passengers. Waiting rooms, ticket offices, the vestibule, and package services were spread out all over the building, and people wanting to transfer had to walk around the front, from the exit to the entrance. For regular people, that is; for some, the situation was much better, as we shall see.

The Görlitzer Bahnhof: Born in War

The history of the Görlitzer Bahnhof is something of an anomaly for Berlin. For over a century, while other stations were built, remodeled, extended, demolished, and rebuilt, became part of Stadt- and other Bahns, or were closed down altogether, little Görlitzer Bahnhof had its own place in Berlin's city life, a somewhat inconspicuous existence in the southeast corner of Kreuzberg. From its opening in 1866 until demolition over a century later, it stood virtually unaltered before gradually disappearing, almost as an afterthought, in what might have been even a greater tragedy than the demolition of the much more famous Anhalter.

Fig. 8.2: Görlitzer Bahnhof shortly after completion, 1867.

Connecting the capital to Görlitz and other cities in the Lausitz region, the westernmost sliver of Silesia, it maintained its own relatively isolated identity regardless of whatever happened around it. Even after the Second World War its front building stood largely undamaged, in contrast to ruins elsewhere. Unlike other stations, it was never confronted with an onslaught of additional services or the explosion of passenger numbers; quite the contrary in fact, because it would even exist without its only long-distance connection (to Görlitz) for some time. However, the post-war desire for car-centric innovation in West Berlin sealed its fate in one of the more egregious cases of cultural barbarism in the city.

The origins of the railway to Görlitz go back to the work of Bethel Henry Strousberg, widely known as the "Railway King." Strousberg made his money in railway speculation and showed an uncanny nose for a good opportunity, which served him well as tensions between the Kingdom of Prussia and the Austrian Empire rose during the mid-1860s. Not only was the city of Görlitz a key point in the connection to the Silesian hinterland, it was also connected by rail to Vienna;

for that reason alone, the Prussian high command saw the line as a crucial strategic addition to their network.

At first, that strategic importance worked against the construction of a rail link to the capital, which had been suggested as early as 1858. At this time, tensions between Prussia and Saxony, an ally of the Austrians, had started to brew, and the Prussian government's demand that the railway be embedded in large and financially crippling fortification works, to prevent misuse by the enemy, worked against the plans. But by the 1860s the mood had changed from defense to offense, making a quick connection to a prospective Austrian or Saxonian foe more desirable.

As relationships with the Austrian empire took on a steadily more grim character, the construction of the railway went into overdrive. By the spring of 1866, the stretch from Berlin to Cottbus had been completed, but construction of the station had not yet started. When mobilization was announced on June 13, soldiers were therefore transported to Cottbus and on to the front from a makeshift platform. By this time, the complex had already been hooked up to the Verbindungsbahn, the ground-level train that connected the Berlin termini to each other to supply them with supplies. However, in this case the *Verbinder* also proved useful to carry troops to the Görlitzer Bahnhof; thus, both became crucial links in a military machine that caused a decisive Prussian victory in the battle of Königgrätz (today Hradec Králové in Czechia) on 3 July 1866. This event not only decided the Austro-Prussian War in Prussia's favor but also immediately proved the railways' strategic importance.

Towering

Despite its crucial contribution to battle victory, there was still no sign of an actual station building on the site. August Orth, Strousberg's house architect (and later the architect of the second Stettin station), had been engaged to design a suitably impressive Berlin bookend to the line. Negotiations to secure a building permit dragged on for months without providing one, but by August of 1866 construction had begun nevertheless. Strousberg had to wait for another year until the necessary paperwork was completed, by which time building had almost been finished; on the last day of 1867, passengers could finally enter the new station to board a train towards Görlitz.

With its distinctive set of towers at the front, separate entrance and exit buildings, and lack of a true terminal platform, Orth's station still reflected some of the elements of the template by Stüler and Strack (both his teachers) which had so heavily shaped the Hamburger Bahnhof (compare Figure 7.2). The station also mirrored the design of its sibling at Görlitz, on the other end of the line, while stylistically it seemed to reference buildings in the Sanssouci palace park in Potsdam, particularly the Peace Church and the recently completed Orangerie Palace.

Fig. 8.3: Inspiration: the Orangerie Palace at Potsdam around 1895.

Like the Hamburger Bahnhof, it contained a turntable for locomotives at the front of a train shed measuring 140 by 37 meters. Even at the time this was a somewhat old-fashioned arrangement; other stations constructed around the same time all had a sliding table (*Schiebebühne*) that allowed for a quicker and more efficient transfer of rolling stock to other tracks. Contrary to the Hamburger, the turntable was obscured from outside view by a building in front of the train shed which, as was still usual, was mainly filled with offices and living quarters for railway personnel.

However, it did contain some passenger facilities, such as a baggage counter and a single waiting room, both of which could be reached through a corridor from the arrivals platform, which was necessary since there was not really a proper arrivals building; instead, an open gallery on the east side allowed alighting passengers to leave the building. Because of a bridge over the nearby Landwehrkanal, the platforms also needed to be built three meters above street level. Combined with the lack of a roof over the engine turntable, this all turned it into a drafty place, especially when arctic gales tortured the city and the station in winter. At such times, passengers at the Görlitzer only left the comfort of their waiting rooms at the last possible instance.

Although many appreciated the monumental station, Orth's design also provoked criticism, which centered on two elements. First was the architectural ambiguity mentioned above, and the fact that the plethora of surrounding building entirely obscured its most important element, the train shed, from view. Secondly, it was felt that this made the building confusing to passengers. Instead of through the prominent front building, they entered through a separate and far less prominent building on the side. Together, these issues gave the impression that the architect had intended the station's core functions to remain hidden.

A Royal Waiting Experience

There was, of course, a good reason that for that impressive front. The whole experience of rail travel was rather different for passengers who happened to be counted among the Very Highest Gentlemen (Allerhöchste Herren) – i.e., the Prussian and German Royals. While royal waiting rooms had existed since the first Potsdamer in 1838, they had never been designed in quite such sumptuous fashion as provided by Orth's new station. Moreover, the royal guests could make use of facilities quite unlike those offered to the common people; a carriage could be driven beneath the arches at the front, enabling a weather-proof exit of the royal (and later imperial) passengers, their guests, and their entourage. Once ready to board the train, they could make their way to it separately, without having to interact with regular passengers.

All this reflected the growth in political self-confidence that the Prussian kingdom had experienced since that first generation of stations. The kingdom had become more important, and so had their rulers; from this point forward this style of royal waiting rooms would become a part of every new station in the city until the First World War.

For the railway company, the royal waiting rooms served two direct purposes, first, to secure the patronage and support of the royal family and the court. Patriotic motives were at play, but also practical ones; all railway companies were getting increasingly dependent upon the state, and for that reason it was important to leave a positive impression on the most powerful persons in the country and their entourage. In addition, they functioned as something of a calling card for both the company and its architect.

Whereas station buildings were regularly subjected to cost-cutting measures during construction, few expenses were spared for these rooms; in fact, railway companies seemed set on out-doing one another in making the royal quarters as op-ulent as possible. Obviously, the general public were not able to use the royal rooms, but they could occasionally tour them at set times and be amazed at their splendor.

The location of the royal waiting room also determined those of the others. In line with Prussian social hierarchy, the first-class waiting quarters were placed the closest, the fourth-class rooms the furthest away; placed further away still, next to the fourth-class waiting room, were the noisy baggage handling rooms.

Fig. 8.4: The platforms at the Görlitzer Bahnhof, c. 1870. The entrance to the Royal Waiting rooms is visible to the left.

Commuters and Coal Traders

After those eventful first months as a military transport hub, the Görlitzer Bahn-hof quickly became a far less exiting place. Over the following decades, its rela-tive isolation gave it a unique status; whereas other stations were being rapidly developed and extended, the Görlitzer mostly turned into a commuter station, in addition to serving the handful of long-distance trains on that single line to Gör-litz and beyond. The first timetable in 1867 contained three daily trains, far fewer than other connections, and that number would not increase by much. After 1883, the station was closed for long-distance trains entirely, which were all directed over the Stadtbahn, with the station relegated to a terminus for local services in the direction of Königs Wusterhausen. They were not restored until 13 years later, in order to relieve the Stadtbahn.

The goods station next to the passenger complex had taken on a far greater prominence in the meantime as it became a crucial point of supply for lignite and building materials from Silesia. This came to determine the character of the quarter around the station as well, with many building firms and coal salesmen setting up shop.

When the Görlitzer opened, the surrounding neighborhood had been designed along with it. At the time, new local legislation regarding construction had just come into effect in the form of the Hobrecht Plan, named after the "planning police's" director, which determined the dimensions of standardized city blocks. As a consequence, the new neighborhood was set up in a grid structure, with streets named after towns next to and served by the line to Görlitz. The square in front was named after the Spreewald, the most common tourist destination served by the station. Set in the easternmost part of Kreuzberg, the presence of small industry quickly gave the area a working-class character, albeit more intermingled with other social classes than the uniformly proletarian Friedrichshain on the other side of the river.

Fig. 8.5: The departures wing, c. 1900.

That also meant that politically, it developed into a decidedly left-leaning neighborhood, and the Görlitzer Bahnhof became a regular spot for socialist and communist leaders to arrive among their cheering sympathizers. Although it rarely took on quite as riotous a character as Friedrichshain, sometimes the inhabitants were prepared to make an exception, as Berlin's Nazi chief Josef Goebbels found out when he had the temerity to traverse the area in 1929:

> I drive a little bit around the Görlitzer station with four comrades, to look at the train and its position from a distance of about fifty meters. While I am unsuspectingly chatting with my fellow party members, the square behind us is gradually filling up. A mob rushes forward. A burly man [. . .] shouts as if he were crazy: "That's the worker murderer Goebbels, workers come on! Now let's put an end to it!" Two hundred to two hundred and fifty men rush forward, armed with lead pipes and clubs. I immediately receive a blow on the shoulder. As I turn to the side, I can just see how one of these animals is aiming a pistol at me. A shot rings out. Then shot after shot! The car is showered by a hail of stones.

The number of trains departing from the station would long remain limited. By 1925, 34 local and eight long-distance trains departed daily, but that number almost doubled during the 1930s. A true second heyday for the station came during another war, although for far more tragic reasons; as other stations were increasingly damaged by bombing, more and more services were relegated to the Görlitzer. From the middle of 1944, however, its own functions were increasingly hampered by wartime destruction and by the end of the war much of the complex had burnt out.

A Slow Demise

By the end of the war, much of the Görlitzer Bahnhof was still standing. The front building had partially burnt out and the roof of the train shed had disappeared, but compared to other stations the devastation seemed manageable. The western entrance building had suffered the heaviest damage, but the first post-war trains were still able to depart as early as late May of 1945. The reason that the station still disappeared owed much to the attitude of the West Berlin authorities and the new order of West Berlin architects, headed by the influential Hans Scharoun, who deemed the building to be "architecturally worthless."

It did take a very long time for the Görlitzer to vanish. First, the train shed went in 1958, and then the entrance building and the rightmost of the two towers. Finally, in 1975, the left-most tower and the arrivals building were demolished. At that time, the area was still used as a goods depot by the GDR train company, the Reichsbahn. Ten years later, goods services had gone as well, and the rails were removed.

Fig. 8.6: Görlitzer Bahnhof in 1954, photographed by Rolf Goetze.

The site saw extensive development after German reunification, when it was turned into a park, which unfortunately can hardly be hailed as a success. Even by Berlin standards, what is now the Görlitzer Park has gained a unique notoriety, most of all as a result of skirmishes between various drug gangs that converge upon the area. The old goods shed is preserved along with an administrative building, and the nearby elevated U-Bahn stop still carries the station's name; together, they serve as a reminder of what was perhaps Berlin's most inconspicuous station.

The Küstriner Bahnhof: The Weird One

About two kilometers to the north and just a few steps from today's Ostbahnhof, the Franz-Mehring-Platz in Friedrichshain is today dominated by the headquarters of *Neues Deutschland*. Once the GDR's state newspaper, it has since been downgraded to a marginal voice in the Berlin media landscape. Although it is still housed in the building it once entirely occupied, it shares the space with various businesses and a furniture storage facility. The area has fallen upon hard times indeed, and the singularly uncharismatic, entirely too broad street does not make things better.

Fig. 8.7: Küstriner Bahnhof, 1871.

Once, however, this was a lively neighborhood that housed a massive railway station on that same square, then called the Küstriner Platz. No station in Berlin, and possibly the world, has seen a more diverse combination of uses than the Küstriner Bahnhof, just north of today's Ostbahnhof; during its life, it saw service as a station, restaurant, hot air balloon workshop, storage compound, and theatre, sometimes simultaneously.

Yet the history of the building shows how it was an afterthought in more than one sense. To begin with, it was not called the "Küstriner Bahnhof" but rather the Ostbahnhof (Eastern Station). However, when it opened in 1867, all of the other stations in the city were named after the destinations they served, and it made sense for people to refer to the new one in the same way, all the more so since it was built on the Küstriner Platz (Küstrin Square), and the name "Ostbahnhof" had also been used colloquially to refer to the Frankfurter Bahnhof (as it had long been the only station in the east of the city) a few hundred meters to the southwest.

Stop-gap

The station was built for two reasons: first, as the Berlin bookend of the Ost-bahn (hence the name), Prussia's first state-built rail line from Berlin, via Küstrin (today in Poland), to Dantzig (today Gdańsk) and on to Königsberg (now Russian Kaliningrad). This was an important line for the Prussian state, and it needed an important station. The original intention had been to connect the line to the existing Frankfurter Bahnhof after its planned redevelopment, but the size restrictions placed on that station's new building made it unsuitable for the long trains running on the Ostbahn, and left the state with no option but to construct a separate facility. In addition, it could serve as a backup of sorts during the construction of the Frankfurter. It is remarkable that the option to close the Frankfurter Bahnhof altogether and replace it by the new Ostbahnhof never seems to have come up.

In 1867, after a relatively quick construction, the new station was opened. It looked quite impressive, and its 188-by-37-meter train shed was the largest of the city.[19] From an engineering viewpoint, it was also an impressive achievement, and for the first time offered passengers a travel experience that was not dominated by a battle with the elements like it had been at the Görlitzer.

The reception of the new building was as mixed as it had been for the Görlitzer Bahnhof, and for largely similar reasons. Those criticisms helped railway architects to reconsider their designs, which we can clearly see in the later designs for the Stettiner and Anhalter stations. Simultaneously, many admitted that the building itself was a beautiful one both outside and in. In a number of ways the Küstriner felt much more modern than the Frankfurter Bahnhof which was being constructed 300 meters further. Moreover, its placement at the intersection of several avenues was considered to be an improvement over its sibling.

In a time where Berlin was still at the beginning of its explosive growth and looking to escape its provincial reputation, this was an important consideration. The Küstriner Platz in front offered a kind of display window for the new station, emphasizing its urban splendor. Not that its surroundings were that splendid; Friedrichshain and the Stralauer Viertel in particular were already acquiring their working-class character, and the many cafés and tenements lining the Fruchtstraße from the Küstriner to the Frankfurter stations were particularly notorious.

The train shed also drew criticism in spite of its innovative structure, with a barrel roof built on steel arches. Most of the roof consisted of glass, but the cen-

---

[19] It needs to be said that, at the time, the recently finished Görlitzer was not that much smaller, and three stations of comparable size were under construction: the new Frankfurter and Potsdamer Bahnhof complexes and the entirely new Lehrter station.

Fig. 8.8: Küstriner Platz with the station behind, 1880.

tral part was made out of sheet metal to keep weight down. This might have been a sensible structural consideration, but the optical effect was that it looked far more massive, and rather less elegant, than a roof fully composed of glass panes. In addition, it kept light out the train shed, and was regarded as something of a blot on an otherwise beautiful building.

A more practical issue was that for the average traveler the front building, built out of red-glazed brick, was entirely useless – even more so than it had been at the Görlitzer. It mostly contained administrative services and housing for railway employees and, again, the only travelers that could make use of facilities here were the royals and their entourage. Functionally it therefore consisted of three largely separated buildings for most people.

There might have been a narrow terminal platform, but it did not really serve any purpose because of this setup; it mainly made it possible to access the middle platform, which was used for excursion trains and for personnel to access the transfer table for engines, possibly the biggest innovation the station presented.

Fig. 8.9: The Küstriner's train shed; note the big black sheet metal sun blocker in the top.

A Quick End

For 15 years, the station saw people coming and going on long-distance journeys to the Prussian hinterland. The Küstriner's importance decreased markedly after the new Frankfurter Bahnhof opened in 1871, while in 1882 it came to an end altogether after that station, now re-christened the Schlesischer Bahnhof, was converted from a terminus into a through station and connected to the Stadtbahn. Its capacity was enlarged simultaneously, which made the use of the Küstriner superfluous and impractical.

By now, the Prussian railways possessed one of the largest buildings in the city without really knowing what to do with it. Maintenance was expensive, and attempts to rent it out initially proved unsuccessful. Thankfully, at least the front building had an occupant in the form of the restauranteur Gerhard Oppermann, whose restaurant Zum Ostbahnhof gained a good reputation throughout the city, drawing upper-class gourmets to the proletarian quarter.

Fig. 8.10: Zeppelin above the Küstriner Bahnhof (top middle, right of the Zeppelin) and Schlesischer Bahnhof (top left), 1911.

In 1884, a tenant for the train shed was found in the form of the Trial Station for Tethered Balloons (Versuchsstation für Captivballons), Germany's first airborne army unit. The long shed turned out to be quite useful to construct balloons, with all the accompanying rope work, while the smaller workshops could be housed in what used to be waiting and luggage rooms. While most balloons were developed for military reconnaissance and signaling, some also saw service to gather scientific data: during 65 manned and 29 unmanned flights, they were used to investigate the atmosphere above the planetary boundary layer. The aim of the Trial Station was to develop standardized ways of constructing (military) balloons, which turned out to be a first step towards later airships such as Zeppelins.

The problem was that by this time, the surrounding area had become a densely populated one and, consequently, taking off in huge balloons filled with highly volatile compounds was regarded as less than advisable. The balloons therefore had to be transported to the Tempelhofer Feld in the south of the city (later turned into a fully-fledged airport) to become airborne: a cumbersome procedure. After only two years, the Trial Regiment left the station again for a venue closer to the field. After another period without occupancy, the Red Cross began to use the train shed for storage.

Plaza!

Fig. 8.11: The Plaza briefly after opening, 1930.

By 1910, the building was still used as a Red Cross storage facility, and people started to notice that it was in a bad state. This showed the lack of maintenance, which was made worse by the fact that the station had originally been built on the cheap, using not-quite-grade-A materials.

The restaurant in the front building continued to attract the better crowd of the area – insofar as there was one – but plaster had long ago started to crumble from the walls, thrown-in glass windows were boarded up all over the place, and even the widows of railway staff that lived on the upper floors had started to move out. The once-grand Küstriner Bahnhof was beginning to look as useless as it was.

To make matters even worse, in 1919 it ended up in the middle of the de facto civil war in the engulfed Berlin after the German loss in World War I. Later, in 1926 the restaurant concluded that the area was far too proletarian to provide enough of its high-class clientele, moved out, and was immediately proven right by a series of skirmishes between disgruntled workers and police that managed to do even more damage, both in real estate and reputation, to the already tortured building.

However, ironically, that episode also saved it for another decade and a half. The owner of Charlottenburg's fancy Scala Theatre, Jules Marx, wished to start a similar venture, but one serving working-class audiences. Looking for a suitable venue that also happened to be in the middle of an area where his core audience had been expressing its right to free speech so ardently, he ended up at the Küstriner Bahnhof. Desperate to get the behemoth out of their ledgers, the German Reichsbahn offered him a 25-year lease on very favorable terms; for instance, the theater was allowed to draw power not from the regular mains, but from the railway's electrical system.

Turning the crumbling structure into an entertainment palace also required a sizeable investment. In the astonishingly brief period of about six months, the station was gutted, with a theater tower created in the middle of the former train shed. Interestingly, the rear part of the train shed continued to be used as railway storage space. The front building, now for the first time in its life used as the station's entrance, was restored to its former splendor. The huge hall was covered by a dome that let through light with a reddish tint. Renamed the Plaza, the Küstriner opened on New Year's Eve, 1929. Its program was expressly aimed at lower-income audiences: "the biggest programme with the smallest prices and the highest perfection."

Fig. 8.12: The grand auditorium of the Plaza. View from the stage.

Initially, Marx' commercial gamble appeared to pay off; reviews were good, intake steady, and an increasing number of famous artists seemed prepared to face the admittedly rowdy crowds at the Plaza. Unfortunately, the economic crisis of 1929 and subsequent slump in attendance proved fatal for the venture, and in 1931 the company had to file for bankruptcy. The theatre, suffering from increasing losses, continued under Marx' direction but with a heavily cut program.

Like everything in Germany, however, the Plaza was also heavily impacted by the Nazi power grab of 1933. Marx, a Jew, was soon replaced by a trusted NSDAP member and the Plaza taken over by a regime-friendly organization, with the theatre coming to concentrate on "Jew-Free" operettas, which immediately presented a few problems. Firstly, finding repertoire that no Jewish text writer or composer had contributed to turned out to be rather difficult: some Suppé and Léhar was all that was left, and attempts to develop new works met with little success. In addition, the Plaza's cavernous acoustics proved less than ideal for the performances.

After this experiment ended in 1937, the theater returned to more traditional variety theatre reviews, this time staged by *Kraft durch Freude*, the Nazi Party's leisure activity organization, which would also be the last incumbent of the Plaza. And as time progressed, the program increasingly became a vehicle for Nazi propaganda.

The End

As will come as a surprise to no one, things did not end well. Friedrichshain experienced its first bombardment in 1943, but the station building did not get badly hit until February of 1945. It was the Battle for Berlin that finished it off for good.

Fig. 8.13: The ruins of the Plaza/Küstriner Bahnhof, 1946, former arrival side (compare with Figure 8.11).

The SS appropriated the Plaza as a temporary HQ for Eastern Berlin, turning it into a prime target for Russian artillery, with predictable consequences; by the time the Nazi surrender was announced, nothing but a burnt-out shell remained.

And where the Schlesischer Bahnhof would be resurrected and become a main transport hub for the German Democratic Republic, the Küstriner quickly disappeared altogether; 1949 saw the beginning of demolition, and three years later nothing remained to remind anyone of what had been opened as Berlin's largest railway station a little over 80 years earlier.

Today there is very little that reminds of the area's past. Not only the station but most of the buildings around it were razed in the 1950s to be replaced by anonymous housing blocks. It will take some effort to bring this corner of the city back to its former, lively – if sometimes perhaps *too* lively – self.

9 Lehrter Bahnhof: Triumph and Failure

And there, late in the evening, is the old, honest Lehrter railway station with its coat of voluptuous red, and behind the platform barriers, this whole Germany of millions and billions is piling up: disgruntled officers who have set up their boot jackets here, and inflationary youths with crisp new leather coats and emaciated working-class women whose bitterness and overexcitement could make them jump at the throats of the whole world. – Fritz Reck-Malleczewen, *Sif. Das Weib, das den Mord beging* (1926)

Fig. 9.1: Lehrter Bahnhof (right) flanked by the German Colonial Museum (left) and the "Glass Palace" at the ULAP Exhibition ground (center), 1910–1915. Contemporary colorized photograph.

Of all the lost stations of Berlin, the Lehrter could be called the least successful from one viewpoint, and the most successful from another. Officially it still exists in name, and it definitely still exists in practice. We know it today as Berlin Hauptbahnhof, and it is the busiest and biggest station the city has ever had; closer inspection reveals it to have been even among its more important, but for a few reasons it never caught the public imagination in the way some of its siblings did.

https://doi.org/10.1515/9783111381879-009

Much of the Lehrter's reputation has to do with its surroundings, since the station was not part of a particularly vibrant neighborhood. Squeezed in-between a large exhibition and entertainment area (Universum Landes-Ausstellungs-Park or ULAP), Berlin's most notorious prison (Moabit), an extensive rail yard, and an industrial port (the Humboldthafen), it was never a place to be unless you purposely had to go to the station.

This stood in stark contrast to the busy roads leading to the Anhalter, the sophisticated atmosphere of diplomacy and culture around the Potsdamer, or even the more than slightly seedy but still bustling environs of the Stettiner and Schlesischer. As part of an anonymous neighborhood, smack in the middle of a *nonplace*, it became slightly anonymous itself.

Quick Lines to the West

The Lehrter owed its existence to Prussia's victory in the war against Austria (1866), with the peace settlement including the inclusion of the Kingdom of Hannover into Prussia, which now came to stretch without interruption from Memel (today Klaipeda) in the Baltic to Cologne on the Rhine. The Prussian state and the Magdeburg-Halberstädter Eisenbahngesellschaft took the initiative to build a line to Lehrte, where it could hook up with existing railways in Hannover; on the Berlin side, the line entered the city together with the lines from Hamburg. Over time, this proved to be a crucial connection not only did it tie Prussia to its western provinces, it also made faster travel to Belgium, The Netherlands, Denmark, and France possible – something the French would find out in 1870, when Prussian soldiers managed to reach their border quicker than they had thought possible.

The Hamburger Bahnhof, which had served the Hamburg line since 1847, had long been one of the larger and better equipped train stations in the city, but by the late 1860s it had already become far too small to handle all the traffic from Hamburg, let alone cope with the trains to and from Hannover as well. But instead of razing the Hamburger Bahnhof and replacing it wholesale, as would happen to the Potsdamer and Anhalter stations, it was decided to build an entirely new station across the road from the old one. The reasons for this decision were twofold; firstly, the railyard near the old station was simultaneously cramped and quite busy, including goods services and all sorts of facilities. To rebuild the entire infrastructure would interrupt services for a long time. In addition, the Hamburger could remain in service for the time being to handle post trains, as well as the express passenger trains to Hamburg.

As luck would have it, a large area nearby, just behind the Humboldthafen, had become available and seemed eminently suitable to build an appropriately

Fig. 9.2: The Lehrter Bahnhof and its surroundings, from the air, 1910, from roughly the same angle as the previous image. Note the amount of industrial installations.

grand new station. As we saw earlier, by this time the railroads – and particularly this one – had become a source of pride, something that both architects and railway authorities felt deserved prominent architecture and a prominent position in the city. Although the Lehrter was an entirely new building in a different place, we can consider it a de facto rebuild of the Hamburger, just like the Potsdamer and others were replaced by new buildings a few years later. The old station would remain in service for passenger traffic until 1884.

Grandeur

Opened in 1871, the Lehrter Bahnhof was first of all a grand statement, much more grandiose in style than the Küstriner, Frankfurter, or even Potsdamer stations, which had opened briefly earlier. Its design, by the architects Alfred Lent, Bertold Scholz, and Gottlieb Henri Lapierre, also contrasted markedly with Berlin's other stations, with its façade consisting of stone rather than the glazed tiling that had become the norm in the capital's official architecture; next to the Potsdamer and Frankfurter it looked almost salacious and was clearly influenced

Fig. 9.3: Lehrter Bahnhof, aerial photograph. Note the impressive arrivals wing (left) and the covered gallery for coaches.

much more by French than by Prussian architecture. There were evident points of inspiration from the grand termini of Paris, particularly the recently completed Gare du Nord, designed by Jacques Hittorff, which had opened in 1866. In addition, it made reference to a triumphal arch, perhaps not entirely coincidentally just after the Prussian victory over Austria – it was the subsequent annexation of Hannover following that war that had made the construction of the line and the station possible, after all. Even for the time, its eclectic style was seen as being rather exuberant – too exuberant for many who were used to a more restrained architectural language that was common for public buildings in the city. Despite its grandiose style, the building still adhered to the usual proportions and functions of a Berlin terminal, including the "Berlin Width" of 37,66 meters and separate facilities for arriving and departing passengers.

The Lehrter Bahnhof's front was dominated by the giant triumphal arch, flanked by two columns on either side. Imposing though this part of the building might be it was mainly placed there for display; the real entrance and exit were housed in the two traditional exit and entrance buildings. The departures wing was a relatively modest affair, but the arrivals wing stood out for its imposing arch, mirroring the one in the façade. It was still not possible for the public to enter through the front of the building – that privilege was reserved for the royals, just as it had been at the Görlitzer. Just as with other stations in the city, the entrance and exit remained separate, although it was possible to travel between

Ankunfts-Seite.

Halle.

Droschken-

Abfahrts-Seite.

Posthof.

Trottoir.

1. Bureau-Zimmer. 2. Sitzung-Saal. 3. Lichthof. 4. Bureau-Diener. 5. Direktor-Zimmer. 6. Vorzimmer. 7. Polizei-Retirade. 8. Flur. 9. Polizei. 9. Betrieb-Bureaus. 10. Portier. 11. Gepäck-Ausgabe. 12. Steuer. 13. Ankunft-Vestibül. 14. Reserv.-Handgepäck. 15. Wartesaal. 16. Büffet. 17. Retirade für Herren. 18. Retirade für Damen. 19. Expedition. 20. Post-Pack-Kammer. 21. Eilgut-Ausgabe. 22. Expedition. 23. Steuer. 24. Betrieb-Bureaus. 25. Vorflur. 26. Betrieb-Bureaus. 27. Entrée. 28. Empfang-Zimmer. 29. Zimmer Sr. Majestät. 30. Passage. 31. Toilette. 32. Gefolge und dist. Personen. 33. Herren-Toilette. 34. Damen-Toilette. 35. Damen-Zimmer. 36. Damen-Toilette. 37. Herren-Toilette. 38. Warte-Saal I. Kl. 39. Warte-Saal II. Kl. 40. Restaurateur. 41. Büffet. 42. Warte-Saal III. Kl. 43. Corridor. 44. Neben-Vestibül. 45. Koffer-Träger. 46. Portier. 47. Gepäck-Annahme. 48. Kasse der Gepäckexpedition. 49. Billet-Haus. 50. Abfahrt-Vestibül. 51. Warte-Saal IV. Kl. 52. Wacht-Zimmer. 53. Encartirung 54. Reponirte Akten. 55. Lichthof. 56. Reponirtes Material. 57. Post-Vorsteher. 58. Dekartirung. 59. Vestibül. 60. Brief- und Pack-Kammer. 61. Post-Pack-Kammer. 62. Station-Vorsteher. 63. Telegraphen-Bureaus.

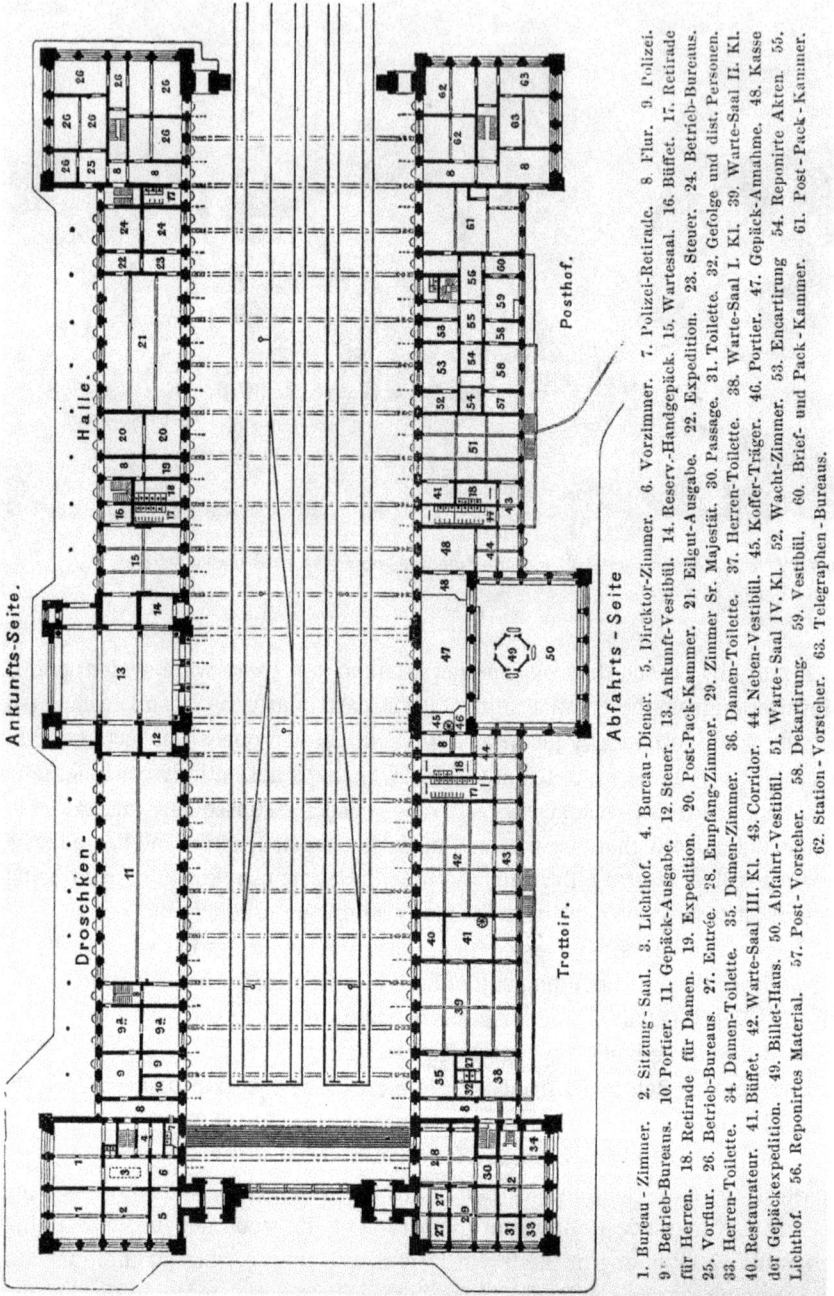

Fig. 9.4: Lehrter Bahnhof, floor plan, original configuration.

the two using an internal platform. In fact, one could walk up some stairs and look out the front windows – but not to enter or leave.

Stylistically, we can regard the building as something of an intermediate between most others of its generation and later ones. The station was a celebration of the Rundbogenstil arches could be found absolutely everywhere. The design did address some criticisms that earlier stations had provoked: there was no useless office building in front, and it made little attempt to hide the fact that it was a railway station as the shape of its large train shed could clearly be seen in but also through the big window at the front. Still, some argued, the square shape of the façade was an obfuscation of the barrel-like roof of the train shed. That window, oriented southward, was also the main source of sunlight for the shed, which was particularly important since, unlike the Küstriner and Potsdamer, it did not possess a glass roof. Windows lined the sides, but they did not let in a lot of light.

The construction of the Stadtbahn in 1882, and the completion of nationalization of the railways in 1884, signaled the end of the Küstriner and Hamburger Bahnhof as passenger hubs, but also impacted the luster of the Lehrter station in two ways. Firstly, the new railroad crossed the Lehrter's tracks immediately behind the station on a viaduct, where the Lehrter Stadtbahnhof was built; while that facilitated easy transfers, it also blocked any sunlight from that side, thus making the hall significantly darker. Over time, this turned out to be a somewhat

Fig. 9.5: The Lehrter's train shed still without the Stadtbahnhof at its end, c. 1875. The big source of light at the end would soon be blocked.

unfortunate arrangement for the Stadtbahnhof as well. Steam from trains leaving the station corroded its foundations, and it needed a thorough refurbishment in 1912 and again in 1926. Ever-increasing traffic using bigger and heavier trains meant that the marshy Berlin soil was starting to give way, causing structural problems for both stations.

Most trains to and from Hannover were now directed over the Stadtbahn, meaning the Lehrter only handled traffic to Hamburg and just a few trains to Lehrte and Hannover, totaling about 22 incoming and outgoing trains a day. With around three quarters of a million travelers annually (1895), that put the Lehrter firmly in last place among Berlin's stations when it came to passenger numbers.

Emigrants and Diplomats

However, the link to Hamburg, the empire's second-biggest city and main port, still was an important one and gave the station not only logistical but also political prominence. Just across the river stood the Alsenviertel; built in a part of the Tiergarten park, this neighborhood contained not only countless embassies but also institutions of the empire, including the headquarters of the royal and imperial general staff and later the Reichstag building.

The proximity of these official institutions turned the Lehrter Bahnhof into a common location for foreign dignitaries to make their entrance into the city, particularly those that had arrived by boat in Bremerhaven or Hamburg. Its representational value was therefore important to the authorities; this explains the unusually monumental arrivals wing and the large square in front of it, where parades and guard inspections might be enacted.

The connection to the Stadtbahn had unforeseen consequences for these state occasions. As the "Gateway to the north," the Lehrter had also become the gateway to the ports of Hamburg and Bremerhaven, from where emigrants made their way to the new world. A steady stream of (generally poor, often Jewish) people and their luggage from the Prussian and Russian hinterland could swell to sizeable numbers, with these throngs often overwhelming the facilities of the elegant station:

> Already in the early 1880s the local press registered growing disorder, especially at the major train stations. Large groups of migrants blocked the limited waiting facilities at Lehrter Bahnhof. Quite a few migrants explored the vicinity of the train stations. Others got into fights or disturbed regular travellers and commuters. In central Berlin neighbourhoods some migrants spent weeks, even months before moving on.

Fig. 9.6: Halls in the reception building of the Lehrter Bahnhof, 1871.

This was a similar issue to the one that had faced the Schlesischer Bahnhof at the other end of the city center, but the important distinction was that in the working-class east, few had really cared about them. At the Lehrter, the obvious poverty and often dubious hygiene of the inhabitants of the waiting rooms clashed rather uncomfortably with the official proceedings. Complaints streamed in, leading to the exit of the emigrants and, in time, to the construction of an entirely new station some distance from the city center, the Emigrants' Station (Auswandererbahnhof) at Ruhleben (see chapter ten).

Fig. 9.7: Former Chancellor Bismarck visiting Berlin, 1890.

With the emigrants mostly gone, the station returned to its more sedate existence, and did not fundamentally change over the following 50 years. It was not particularly busy; in fact, the Lehrter Stadtbahnhof carried considerably more passengers than the "main" station did. The area also never became much more attractive; as posh and upscale as the Alsenviertel might have been, it was not the most dynamic quarter and mostly closed down after working hours. Some life was added at times by events at the ULAP, a large exhibition venue next to the station that hosted alternating events. The vaguely oriental Exhibition Palace was a well-known sight as the venue of various trade fairs and the annual-*ish* Great Berlin Art Exhibition.

However, the immediate vicinity of the station remained dominated by a freight port, the Humboldthafen, and various other industrial installations including a foul-smelling matches factory. The relatively low number of passenger trains meant that the rail yard could be used more intensively for freight, and a new goods station opened in 1881 right next to the arrivals building. However, that also gave the area an even more industrial feel, thereby further diminishing its attractiveness for regular travelers. The massive and rather gloomy complex of Moabit prison around the corner also did little to improve the atmosphere.

Destruction and Revival: The New Lehrter Bahnhof

That did not mean that nothing remarkable took place. On 19 December 1932, a new rapid train, the streamlined Fliegender Hamburger (or "Flying Hamburger"), was first used on the Berlin to Hamburg run, setting speed records and establishing the fastest train connection in the world at 160 kilometers per hour. More fundamental developments were in store, however, and as usual during the Second World War, they set up the station's demise. From 1944 onwards, it proved to be an easy target during air raids; various parts of the station, including the train shed, had burned out by the end of hostilities.

Damaged but not irretrievably so after World War II, it remained in use for a few years after 1945 and could perhaps have served longer, because unlike many other terminal stations its tracks were still used to connect Berlin to Hamburg. Both architecturally and logistically worthy of salvation, it nonetheless fell victim to West Berlin's relentless drive towards modernization; few even noticed when it got blown up in 1959. In a final bout of ignominy, a movie news report about the destruction of the station even mixed it up with the Görlitzer.

Because of its proximity to the Berlin Wall, the area remained mostly unoccupied for the subsequent three decades, with the Stadtbahnhof remaining as a forlorn memento of what had been. That all changed in the late 1990s, when the site was chosen by the German state and Bundesbahn as the location of a shiny new Hauptbahnhof, the main station for Germany's new capital.

With the Stadbahn now used for regular train travel again, the Lehrter's location opened the possibility of creating, for the first time in Berlin's history, a true train hub, combining the Stadtbahn with connections to the north and south, this time through a new tunnel. A large new station was opened in more or less the same location as the old one in 2006; initially, it was even given the weird hybrid

Fig. 9.8: Family in the ruin of unused the disused Lehrter Bahnhof, 1958.

name "Hauptbahnhof / Lehrter Bahnhof," but nowadays the old name is only left at the Stadbahn platform – although few people are probably even aware of it.[20]

Unfortunately, the construction of such a huge new station meant sacrificing the Lehrter Stadtbahnhof. The Deutsche Bahn, at the time run by the almost proudly unsentimental Hartmut Mehdorn, justified this with the argument that

---

20 It's even a bit more complicated. The subterranean station, where the tracks have the same basic orientation as in the old station, is officially still called Berlin Hauptbahnhof - Lehrter Bahnhof (abbreviated BL – a designation only used internally); the Stadtbahn tracks for local and long-distance trains carry the designation Berlin Hauptbahnhof - Lehrter Bahnhof (Stadtbahnhof) (abbreviated BLS).

Fig. 9.9: The Lehrter Stadtbahnhof shortly before demolition, 2002. To the right, the canopy of the new Hauptbahnhof can just be made out.

there were still plenty of these stations around from the same era of the Stadtbahn, such as the Hackescher Markt and Bellevue. Make of that reasoning what you will.

Whether you like the new Hauptbahnhof or not (I'm torn, but we will return to that in chapter eleven), there is no doubt that it is a success. For one thing, its 120 million annual visitors are testimony to that – almost double the amount of all of Berlin's stations in 1894 combined. It is truly a Hauptbahnhof in a practical sense, connecting railways from all sides. Little has improved in the neighborhood, though; one only really visits the area around Hauptbahnhof to go to Hauptbahnhof. Thankfully, it is now connected far better to the city's transport infrastructure than the old station ever was.

The Lehrter's Children

As unlikely as it may sound, the original Lehrter Bahnhof can still be seen in action – almost. There is a fairly exact facsimile of the building, located in Budapest; called Keleti Pályaudvar (or Eastern Railway Station), it comes very close to being a full-on duplicate of the Lehrter. The building was designed (using some tracing paper and a pencil, one assumes) by the architects Gyula Rochlitz and János Feke-

Fig. 9.10: Keleti Railway Station, Budapest, ca. 1890. Photographed by Karl Löhle.

teházy and opened in 1884 to serve train lines to Transylvania and the Balkans. This was made easier by the publication of an extensive article about the Lehrter in an architectural journal, the *Deutsche Bauzeitung*, which included various technical details and drawings, including a ground plan (See Figure 9.4).

While there are some stylistic differences, these are minor; it has almost the same width (42 vs the standard 37,66 meter Berliner Breite for the Lehrter) and height, the same layout (including, originally, entrance and exit buildings and track layout) and an almost identical façade.[21] It was restored to much of its original splendor during an overhaul of the area in 2004, which means that anyone willing to travel to Budapest can experience something very close to the Lehrter Bahnhof for themselves, even 70 years after it has gone.

While a visitor to Keleti might look upon it as an example of pure plagiarism, it should be remembered that many railway architects habitually took inspiration from examples both home and abroad. Particularly in the early years of the railways,

---

21 Although Keleti sports a public front entrance today, that is a later adaptation. The Lehrter's train shed was about 40 meters longer, although the platforms are about the same lengths; in Budapest they consequently extend further into the open air.

when a consensus on what a station should look like was still up for debate, architects and engineers were on the lookout for well thought-out designs to base their own work on. For instance, the published drawings and situation sketch for Berlin's Frankfurter Bahnhof (1841) served as model for Amsterdam's Rhijnspoor station of 1843, while that city's innovative Willemspoort station (1843) in turn looks suspiciously similar to Paris's slightly later Gare d'Enfer (today Denfert-Rochereau, 1847).

Station architects clearly habitually kept their eye on previous designs to inspire them. This was not just a matter of aesthetics; technological innovations also needed to incorporated in new structures. An example of such a development are the various ways devised to get trains from one track onto another inside the station. In a terminus, it is important to get the pulling locomotive, which has literally reached the end of the line, out of the way and ready for new use; initially, turntables were used to achieve this. A locomotive would be driven on, then turned around and driven away, but this method had two disadvantages: it took up a lot of space and the diameter of the turntable restricted the length of rolling stock. Soon, locomotives started to become too large for the table. The replacement was a shifting table, where a locomotive would drive onto a bit of track on a cart, which would simply be pushed from one track to the next. This still was not ideal, though; it took time and quite a bit of manpower or even dedicated steam power, as locomotives became even more massive. The final innovation was a set of points inside the train shed, which required space also, and combined with the increasing length of trains it required ever-longer sheds.

Each of these solutions was essentially hard-wired into the buildings that contained them, and conversion to other ones could be costly and sometimes impossible. For that reason alone, railway engineers and architects needed to keep abreast of recent developments, but they also kept their eye on other designs. As we saw in the case of the second Stettiner Bahnhof, for instance (see chapter four), its design was at least partly influenced by Paris's Gare de l'Est. The Lehrter itself was no different; the combination of a central arch with two vertical columnlike elements which made up its façade could be traced back to at least to Braunschweig's main station of 1848. Although radical solutions were sometimes implemented by choice or necessity the norm was usually more iterative, building on earlier examples. Even so, Budapest's station probably crosses the line between inspiration and plagiarism.

The Lehrter's influence wasn't limited to Budapest. In 1879, the university city of Leiden in the Netherlands got its second, much grander railway station; designed by Dirk Margadant, something of a specialist, it took clear cues from the design of the Lehrter. A few decades later, Margadant designed the much better-known Haarlem station, which opened in 1908. Still in use today, it is widely considered to be among the Netherlands' most beautiful stations and an example of

Fig. 9.11: Leiden II shortly after opening, 1870.

Fig. 9.12: Eastern (main) entrance of Haarlem Station (1908).

art nouveau design in railway architecture. But some of the Lehrter Bahnhof's DNA returns to Margadant's work in the form of a monumental entrance, with a large, arched window flanked by two pillar-like structures, with those rounded forms repeated to either side.

The Lehrter Bahnhof itself might not have left the biggest of most favorable impression on its contemporaries in Berlin, but when it comes to spawning an architectural legacy, we may perhaps consider the Lehrter Bahnhof to have been the most successful of them all. And considering what happens on the site today, even the most important.

10 The Beginning And The End of the Line: Berlin's Dedicated Termini

So far, we have been looking at places intended for general passenger travel. However, the city contained multiple other rail installations; some were for cargo traffic of various kinds, ranging from beer barrels to cattle to mail, while others transported passengers to certain activities, such as race tracks or other sports venues. And finally, there were stations that were intended for specific groups of people, some of whom were dedicated to removing people from the city permanently. Something else they shared was that, for several reasons, these facilities were placed away from the public gaze, in the suburbs or sattelites of the city.

Getting People to the New World

Fig. 10.1: At the Emigrants' Station, Ruhleben, 1895. Drawing by Werner Zehme.

Perhaps the most innocuous, relatively speaking, of these stations was a facility opened in 1893 for the purpose of transferring emigrants out of Germany on to the New World (mostly the United States). For decades, prospective emigrants,

https://doi.org/10.1515/9783111381879-010

many of whom came from the eastern Prussian lands, had already used the capital as a point of transfer to trains that would bring them to the ports of Bremen or Hamburg, and eventually to the New World. This was really unavoidable, since hardly any through services existed. Initially most of that stream had grouped together near the Schlesischer Bahnhof in the east of the city and, although totally overwhelming its cramped facilities at times, the authorities did not really devote much attention to the problem. After all, these people were usually from the lower classes, often poor Jews and Poles, and so were considered to fit in perfectly well with the equally proletarian spirit of Friedrichshain. The east of the city was considered to be something of a social wastebasket, and not as worthy of official attention.

Things changed dramatically, however, once these migrants increasingly descended on the Lehrter Bahnhof after 1882. Originally the point of departure for Hannover, nationalization and rationalization now turned the Lehrter into the main station to Hamburg and on to Bremen. Moreover, the newly opened Stadtbahn now made it much easier to travel onward from the Schlesicher Bahnhof.

The authorities had clearly not thought the consequences through, but they also did not have luck on their side. From the middle of 1881 onwards, the western portion of the Russian Empire was wracked by semi-official pogroms after the violent death of Tsar Alexander II, which in well-established Russian tradition was blamed on "foreign agents," usually Jews. The anti-Jewish waves of violence and other forms of official repression not only gave a first impulse to Zionism but also created a fresh rush of Jewish emigration towards overseas destinations via Berlin, Germany's main rail hub – in addition to an already sizable presence of "regular" emigrants. Within no time at all, the Lehrter Bahnhof turned into a virtual refugee camp, with as many as half a million people a year waiting, sometimes for weeks, to catch a connection to the boats.

This time, the authorities minded very much: the Lehrter was a much more central station. Its position right next to the posh Alsenviertel had made it a regular point of arrival for official guests, but it lost too much of its representative value when these had to compete with an overabundance of migrants. Apart from social and political considerations, there were also very real concerns for health and safety; many of these Russian immigrants spoke little if any German, had no source of income while camping at the station, and their sanitary and health states left much to be desired after the long journey.

As the stream of migrants continued to increase, by the mid-1880s it had become clear that the situation needed a more satisfactory and permanent solution. In the middle of an international cholera epidemic, the understandable public fear of contagious illnesses, combined with worries about crime, caused an outcry which led to the creation of a temporary facility in one of the arches of the Stadtbahn near Charlottenburg Station. Although mostly out of the public eye, this solution still left much to be desired as the space was also cramped, dark, and badly ventilated, and therefore hardly addressed the hazard of disease. Moreover, within Charlottenburg, "Germany's richest city," the measure proved predictably unpopular.

Soon, construction was initiated of a dedicated facility further away still from the city, in the fields between Spandau and Berlin's western satellite cities. On 11 November 1891, this new station, specifically designed to process emigrants, admitted its first "guests." Although perhaps more suited to the requirements of its inhabitants than the Stadtbahn arches, the Auswandererbahnhof Ruhleben (Ruhleben Emigrant Station) was a far cry from the elegant surroundings of the Lehrter Bahnhof. Built in the traditional mock-half-timbered style reserved for temporary structures, it felt less like a station than like an internment camp.

Fig. 10.2: Disinfection of new arrivals at the Emigrant's Station in Ruhleben. Drawing by Werner Zehme.

New arrivals would first be checked for papers, including a boarding ticket, and financial means. If they lacked the means to sustain themselves or pay for a return trip in case of rejection by their destination country, they were to be sent back. Then, they were cleaned and disinfected before being allowed to await further transport in one of the station's three barracks, each of which was designed for about 200 people and constructed out of corrugated iron, clad in wood. More stringent medical oversight was added in 1892, after a large cholera epidemic had caused havoc in Hamburg, the main destination port; both Hamburg and Bremen threatened to block outsiders from entering unless they could show a doctor's certificate of health. A medical compound was added to the complex to prevent Berlin from getting stuck with all the emigrants; during its heyday, the facilities and the station processed over a million people a year.

The hastily built compound soon came to be regarded as something of an eyesore that was getting increasingly noticed as Berlin and its satellites expanded. Plans to a build a horse racing track near the site led to an intention to move it – and its unloved inhabitants – even further westward, out of the urban area. However, in the end it was not so much the horses as the First World War that put an end to the Auswandererbahnhof; as hostilities between the German and Russian empires were initiated, the stream of emigrants dried up overnight, removing the need for a dedicated station. Although plans for new facilities were drawn up during the war, nothing came of them, and after the conflict had ended migration numbers never reached the same levels again, also because of tightened immigration restrictions in the United States.

Fig. 10.3: A surviving barracks (now demolished) at the Emigrant's Station in Ruhleben, early 2000s.

The remains of the station were allowed to be left standing until the beginning of the twenty-first century. Despite their protected status, they were removed in 2012 because no use for them could be found; this way, Berlin's bookend of the line leading to New York's famous Ellis Island disappeared with a casualness it really did not deserve. Today, the site is occupied by an industrial estate.

Learning to Conquer the Railways: Schöneberg's Military Station

Fig. 10.4: Schöneberg Military Station around 1910.

The military station in Schöneberg is something of an odd one out on this list, because it certainly was not intended to give people a permanent farewell; in practice, however, that was often the case. Prussia had realized the military potential of the railways early on, and they had played a pivotal role in its victories against Austria in 1866 and France in 1870. Particularly that last war, which had led to the founding of the German Empire itself, led to the construction of an elaborate system of military railways around the capital – much of it paid for by French war reparations. The core of this system was a dedicated connection built immediately after the war and leading from Berlin across the country to the Lotharingian city of Metz, recently annexed from France. Somewhat irreverently dubbed the *Kanonenbahn* (Cannon Railway) by the public, this railway made it possible to rapidly deploy troops and material to the western borders of the empire. Much of the line remained off-limits to regular passenger traffic, and this certainly applied to parts in Berlin; the whole purpose was to avoid the risk of using the congested regular networks. Berlin's first section, from the new Grunewald station on its western edge to Wetzlar, was inaugurated in 1879.

However, Berlin's first own military railway opened well before that. Immediately after the victory over Austria in 1866, a battalion of "Railway Pioneers" had been created after an American example; a training ground was chosen in Schöneberg, which was then an independent city located to the south of Berlin and adjacent to one of the city's main parade and exercise grounds. Initially, a dedicated railway was planned to connect the installations to the shooting range at Tegel, but that plan was cancelled once the range was relocated to Kummersdorf, situated to the south of Berlin. Luckily, this new location happened to be very close to a newly planned railway line connecting Berlin to Dresden, which faced some challenges in obtaining the necessary permits. That railway company now found the Prussian War Ministry on its side and, in exchange for the elimination of bureaucratic obstacles, a military line was to be created running parallel to the Dresden railway until Zossen, 30 kilometers south of Berlin, where it diverged towards Kummersdorf and the extensive exercise grounds at Jüterbog. The railway company paid the construction costs of the line, but in return the military authorities offered them a good deal for the site of the prospective Dresdener Bahnhof (see chapter 3).

The decision to construct the Royal Military Railway was not merely based on the wish to transport troops and artillery to training facilities. While soldiers could be transported out of the city, they could also be brought into it if the circumstances called for it, the civil takeover of Berlin during the revolution of 1848 still being a present memory. However, the main purpose of the Berlin complex itself was not to handle passengers, but rather a training facility for the instruction of the railway regiment on how to build, destroy, and manage railways in wartime situations. For that reason, a fully functional station building, several sheds and workshops, and a depot for the railway battalion's field equipment were added. The Railway Brigade mainly saw action in the German colonies, constructing bridges in German East Africa (Tanzania) and West Africa (Namibia) and rebuilding the Chinese railways after the Boxer Uprising of 1900.

The erection of the building and its adjacent facilities was started in late 1874 by the railway battalion itself after it had been housed in barracks on the site. This had been a requirement from the War Ministry, which saw the task as a good opportunity for the unit to prove itself and make a saving on the expense at the same time. The relatively modest station building was designed and built in a medieval revival style more reminiscent of other Prussian barracks than the city's other railway stations, with eagles adorning the four corners of its two towers.

A few months after the Berlin-Dresden Railway opened in 1875, the military railway line running next to it became operational. Initially, the line was used mainly to carry troops and heavy artillery to and from the shooting ground at Kummersdorf. Starting from 1888, and after public and political demands, the military station was also made accessible to civilian travelers; people could travel from Schöneberg train station to Jüterbog along the military railway, and at a relatively low cost. It never became a significant connection, however, with usually less than a dozen departures per day in either direction. The long, straight stretch of railway also allowed for other applications; from 1901 onwards, Siemens and AEG undertook speed record attempts on the single-track line in order to test the possibilities of electrification.

Fig. 10.5: Station shortly before demolition, 1955.

Because the installations at Schöneberg symbolized Prussia's and Germany's offensive approach to war, Germany's defeat in the First World War, and the consequent dismantling of the Prussian army after the Treaty of Versailles, rendered them and the station mostly obsolete. The entire Military Railway was turned over to civilian authorities, but the station itself – now renamed after the nearby Kolonnenstraße – was too awkwardly situated in the city to be of much use to passengers in the long run; it was therefore closed for passenger services in 1920, and much of the line dismantled. After all, the Berlin to Dresden railway ran along the same route and could take over most services. Only the stretch from Zossen to Jüterbog – now effectively a branch line – was kept open. From that point onward, the area, conveniently situated immediately south of the Anhalter Bahnhof's goods facilities, was exclusively used for the handling of freight.

The station building itself began to decay, like many other Prussian military buildings. Abandoned and heavily damaged during World War II, it was eventually demolished in 1955. Little by little, the sizable compound at Schöneberg was used for other purposes until little remained as a reminder of Berlin's military railway installations. Only a few administrative buildings survive, which have been repurposed as offices and apartments.

Grieving the Dead: Berlin's Funeral Line

Fig. 10.6: Stahnsdorf Station shortly after opening in 1915.

One of the challenges facing the rapidly expanding capital in the nineteenth century was what to do with its deceased. As the population continued to balloon, the existing funerary facilities proved entirely insufficient to deal with the predictably growing number of dead Berliners. As early as 1878, a report estimated that the cemeteries within the city walls would soon be full; initially, the protestant church had been tasked with supplying burial grounds, but in 1890 that responsibility fell to the city.

Initially it attempted to tackle the problem by opening several smaller new cemeteries outside of the city limits, but the rate at which these were filled made it clear that more drastic measures were needed. After some back-and-forth, it was decided to construct three large cemeteries for the city. To serve the southwest of Berlin, a suitable area of one and a half square kilometers was found some distance from the city, two kilometers west of the village of Stahnsdorf. Purchased in 1902 and taken into service seven years later, it became the main funerary facility for the southwestern part of the city.

From the beginning, the construction of a dedicated railway branch to carry both deceased citizens and grieving relatives to the funerals (and the latter category back as well) was part of the plans. Stahnsdorf might have possessed a tram connection to the city, but its capacity was considered insufficient for the many visitors expected at the burial grounds. The distance – of a short five kilometers – to

Fig. 10.7: Diagram of the railways around Wannsee and Potsdam, 1914 by Mark Thomas.

the nearby suburban station of Wannsee potentially allowed convenient connections to the inner city.

There were some technical issues to be tackled, most notably a bridge across the heavily used Teltow Canal and a crossing of the *Stammbahn*, Berlin's oldest railway. However, the main hurdle proved to be organizational; although a basic agreement about the construction of the line was reached in 1908, it took another five years for it to be completed, due to various disputes between the Evangelical Synod, the city, various landowners, and the Royal Railway Directorate. But on 3 June 1913, the station *Stahnsdorf Friedhof* (Stahnsdorf Cemetary) could finally be opened, initially with an hourly passenger service; from Wannsee, the journey took eight minutes. In addition, a single daily train carried the corpses that were to be buried to the facility from Halensee station, where they had been collected and stored.

In Stahnsdorf, the grievers arrived at a station that was modest both in style and dimensions; much like the military station did at Schöneberg, its design also revealed something about the reason for its existence. In this case, the station resembled a funeral home as much as it did an urban railway station. This was hardly coincidental; its designer, *Baurat* Gustav Werner, had also been responsible for the central mourning chapel of the cemetery as well as other funerary structures.

Positioned some distance away from the road, the building possessed both goods (for the deceased) and passenger (for the living) amenities. It contained some special facilities to handle corpses, such as a cooler and a special hall some 300 meters away from the passenger station that allowed for discreet removal from the train.

From the beginning, plans were developed to extend the small and little-used railway line into something more useful, and extending the railway further south to the town of Teltow and then on to the main S-Bahn network. Another scheme involved the construction of a large Jewish cemetery near Dreilinden, the only other stop on the line. Predictably, post-World War One conditions put these plans on ice. A much more ambitious plan launched in 1936, and involved a total rebuild of the railway and station as part of Hitler and Speer's *Welthauptstadt Germania* (see chapter eleven). Like most things about that project, nothing was ever realized. However, when the rest of the S-Bahn network was electrified in 1928, the funeral line wasn't forgotten.

That was it, though. In the end, the little station at Stahnsdorf remained nearly unchanged during its life; the most significant alteration was perhaps a renaming into *Stahnsdorf Reichsbahn*. It saw daily throngs of mourners shuffle in and out, without significant alterations. Even the Second World War only brought a temporary diversion when the bridge across the Teltow Canal was blown up in May 1945; it took about four months for service to be resumed to Dreilinden, and from 1948 onwards passengers could again reach the cemeteries like they had before the war.

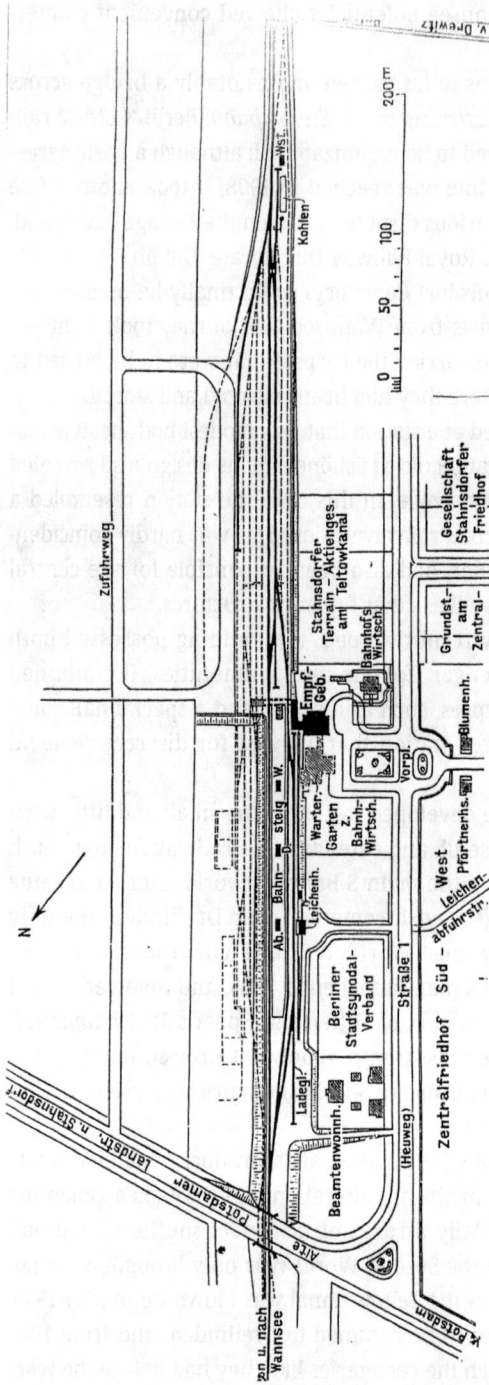

Fig. 10.8: Map of the installations in Stahnsdorf.

The Cold War finally offered an obstacle that the Friedhofsbahn was quite liter-ally unable to overcome. Since Stahnsdorf was located in the Soviet occupation zone, and Wannsee in the American zone, passengers were subject to passport controls. An early attempt by East German authorities to quietly kill the line in 1953 failed because of pressure from the church, but such scruples were entirely thrust aside with the construction of the Berlin Wall on 13 August 1961. Since it ran directly north of Dreilinden, the Wall separated the line's only two stations from the hub at Wannsee. For West Berliners, it became practically impossible to be buried at the cemeteries or even visit them.

The station building, now useless, served various functions after its closure. First, it became the home of a state-owned agricultural company, then a storage space for textile products. The platform was stripped of its roofs and other struc-tures, which were then used to rebuild Warschauer Straße station in the east of the city. In 1976, the derelict station building was finally removed.

Attentive walkers in the forest may notice remains of the line here and there; however, little by little even these disappear. Some of the station's platforms re-main, although they are overgrown; a small nearby monument is as modest as the station it depicts once was, and offers some explanation. Occasional attempts to imbue the Friedhofsbahn with new life have consistently been discouraged by Deutsche Bahn, and it looks as though it is gone forever.

The Railway to Genocide: Platform 17

We have left what is in one way Berlin's most sinister station for last, and it seems eerily fitting to discuss it after the funeral line. Today, Grunewald station languishes in relative obscurity in a prosperous western suburb. The beginnings of the station were relatively innocent; constructed in 1873, initially it only served the Kanonenbahn, the military railway to France. A first, very basic station build-ing was opened in 1879 as Berlin Hundekehle; possessing four platforms, it also offered a connection to Berlin's circular railway, the Ringbahn. Five years later it was renamed Grunewald – its present name – after the forest next to it.

By the end of the nineteenth century, the area was being developed into a garden city for the better-to-do. The railway was also increasingly used for civil traffic, and it needed a proper station that reflected the neighborhood's elevated status. The railway architect Karl Cornelius had just designed the renovation of the Stettiner Bahnhof (see chapter 4) and was engaged to create a properly im-pressive building; opened in 1899, it used a pastoral style to reflect the idyll of the surrounding suburb, referencing a castle gate.

Fig. 10.9: Deathly destination: Grunewald's Platform 17.

In addition to trains serving the Ringbahn, from 1928 onwards Grunewald also served the electrified S-Bahn; that, and its military legacy, caused it to become unusually large for a suburban station. Unfortunately, that surplus capacity also turned it into a convenient location for the mass transport of Berlin's Jews and other "undesirables" to the concentration camps during the Holocaust. From 18 October 1941 onwards, 186 trains departed from platform 17 at Grunewald station for the concentration camps.

Jewish citizens were shipped off from other stations as well, most notably the Anhalter Bahnhof. The big difference was that at the Anhalter, this happened in relatively normal circumstances, using the D-train carriages commonly in service with the Reichsbahn, all such pretense got cast aside at Grunewald. Here, people were unceremoniously pushed into cattle cars to begin their last journey. Comparable to the Auswandererbahnhof before it, for the authorities Grunewald had the advantage of being out of the way, and therefore out of public sight.

Before the rise of the National Socialists, Berlin's Jewish community had stood at around 160,000, the largest in Germany. Half of that number had left if they could afford it and were allowed to in the pre-war years of Nazi persecution. During the War itself, 60,000 of them were transported away from the capital, initially to ghettos in Łódź, Warsaw and Riga, later mainly to Auschwitz-

Fig. 10.10: Memorial plaques commemorating the transports from Grunewald to the death camps.

Birkenau. Around 50,000 of those people started their death journey at Grune-wald, and as we know very few returned. Many did not even reach their destination. Berlin's Jews would first be collected at various points in the city: in synagogues, hospitals, old people's homes or even at cemeteries. Faith in a good outcome was predictably and justifiably low, and suicides were common.

From these gathering points, they were brought to the station. Even if they had wanted to, they would have been unable to admire Cornelius' entrance building; rather, they were driven directly onto the cargo platform 17 using its access ramp, crammed into cattle cars and driven away. Of course, the cars were entirely unsuited for human transport, and usually too full; people had to stand during the entire, two-day journey, urinating and defecating inside the car. Many perished well before reaching their destination, and even more immediately after, in the gas chambers of Auschwitz.

Obviously, aside from the platform itself little remains of the terrible scenes that took place at the station. After the War, the authorities took their time before installing some fitting form of monument to its role in the Holocaust. Community-supported memorials were placed but tended to disappear soon after. Eventually, a monument designed by the Polish artist Karol Brionatowski was unveiled in 1991, next to the ramp to platform 17; it shows the silhouettes of people against a concrete wall. Some time later the Deutsche Bahn – the successor to the Deutsche Reichsbahn Gesellschaft that dutifully carried out the deportations – added iron plates on the tracks detailing each of the trains that took people to be murdered in the camps.

Despite these efforts Platform 17 still lies there somewhat forlorn. Possibly the most visible reminders of the Holocaust are several birch trees taken from Auschwitz-Birkenau and planted in front of the station in 2012. Inside Cornelius' building, as pretty as ever and remarkably free of graffiti by Berlin standards, the everyday commuter can pass these various commemorations without really noticing them; that, if anything, is perhaps an unwitting but effective reminder of the banality of evil. It may get even more banal; plans are in development for new apartments next to the site, which would necessitate removing both the ramp to platform 17 and Brionatowski's monument.

11 Plans! The Long Road to Hauptbahnhof

> The coach dropped them back at its pick-up point outside the Berlin-Gotenland railway station. [. . .] The entrance to the station was disgorging people; soldiers with kitbags walking with girlfriends and wives, foreign workers with cardboard suitcases and shabby bundles tied with string, settlers emerging after two days' travelling from the Steppes, staring in shock at the lights and the crowds. Uniforms were everywhere. [. . .] From here, trains as high as houses, with a gauge of four metres, left for the outposts of the German Empire – . . . It was the terminus of a new world. Announcements of arrivals and departures punctuated the 'Coriolan Overture' on the public address system. – Robert Harris, *Fatherland* (1981)

Almost from the beginning, there was profound unease with the way Berlin's railscape was developing. Part of the reason for this was a lack of coordination between the various railway companies, finally addressed (but never entirely solved) with nationalization in the 1880s; since then, several scenarios – some far-reaching, others more iterative – were proposed to cope with the steady increase in demand for rail services, and to solve issues of fragmentation that were present from the beginning.

Whatever one thinks of Berlin's Hauptbahnhof, it is still the hub of arguably the most comprehensive railway system the city has ever had. To get there, however, took about 150 years and countless proposals. This chapter is about those attempts to turn Berlin into an easier city to get to and to leave, but most of all to fundamentally change its character from a city to arrive in or depart from to a city that could be easily traversed.

The railway infrastructure of the Prussian and later German capital was a hotly debated topic and many more ideas have been discussed than are mentioned here. Nevertheless, as much as the proposed solutions diverged, we will see that they also arrived at a basic form of consensus relatively quickly.

The First Hauptbahnhof

One of the earliest, and arguably the most intriguing plan for a central railway station for Berlin, was devised in 1865 by Ludwig Carl Scabell, the first chief of Berlin's firefighting force and something of an innovator in urban services. He had to; the breakneck pace of construction going on in the rapidly expanding capital did not always include proper safety precautions, and a decent water and firefighting infrastructure was essential to prevent disaster.

https://doi.org/10.1515/9783111381879-011

As early as 1865, Scabell proposed a large central station immediately behind the University building (now the Humboldt University) at the Kupfergraben, opposite today's Pergamon Museum. Scabell envisioned a raised station building and railways to prevent interference with regular street-level traffic and shipping, and the proximity of the Spree also made it possible to combine the station with a freight harbor. It was an ambitious idea, which might have solved Berlin's transport problems for the foreseeable future.

Fig. 11.1: Ludwig Scabell, proposal for a central station (1865).

Of course, it was also going to be hugely expensive because of its size and location, and potentially problematic because it would force several independent companies to use the same facility. This was a problem, because not only did these companies not always see eye to eye but around 1865 no fewer than four of Berlin's stations were either under construction or in advanced stages of planning. They would still need to be finished, since Scabell's station could take a decade to complete, and the idea of four brand new stations being closed after a handful of years of service was too much. Scabell's concept was quietly shelved but never entirely forgotten.

However, it was not the end of Scabell's involvement with the rationalization of Berlin's railways, since he led the conversion of the Schlesischer Bahnhof from a terminus into a through station when it was hooked up to the Stadtbahn in 1882. The Stadtbahn, or rather the combination of the Stadtbahn and the Ringbahn (see chapter 6), can be said to have been the city's most comprehensive plan at reforming its railway infrastructure.

While the Stadtbahn was broadly successful, it also contained some short-comings, two of which stuck out in particular. Firstly, instead of connecting the old termini it added several new stations of its own, which contributed to the chaos because the added flexibility of the system meant that trains from a destination might now be relayed to one of several stations in Berlin. Previously, one at least knew where to go for a given destination, and while it provided a hugely practical service for east-west travel throughout the city, it did little to improve north-south connections. Arriving at the Anhalter Bahnhof in the south, the quickest way to get to the Stettiner Bahnhof in the north was still to get a horse-drawn cab. It was certainly possible to catch a train, but the circuitous route along the Ringbahn hardly saved much time.

Contests

The discussion about the reform of Berlin's railway infrastructure took on a new urgency around the turn of the twentieth century as part of broader discussions about the future of the capital. Already in the early nineteenth century, various ideas about the amalgamation of the city into a Greater Berlin had been circulated, but by the end of the century these had gained a greater urgency. To begin with, satellite cities such as Charlottenburg and Schöneberg had become closely integrated to Berlin proper, sharing services and transport, and their administrative independence came at the price of much-needed co-ordination of urban development. In addition, it was a politically attractive idea, since it would create a capital worthy of one of the world's great empires. Of course, such a city required a tightly integrated transit system to unite that capital.

For the Schinkel architecture contest for 1897, a design was requested for a north-south rail connection. The winner, Gustav Schimpff, conceived an elevated north-south connection across the city, to connect the Potsdamer Bahnhof to the Stadtbahn at the Lehrter Bahnhof and then on to the Ringbahn in the north. The plan was clearly inspired by the Stadbahn and involved a circuitous route across Potsdamer Platz to connect to the Potsdamer Ringbahnhof. However, an underground connection via Friedrichstraße station was also discussed as a solution and would become a favorite idea.

A rather more comprehensive approach was taken a few years later in 1908, when the Berlin Architects' Association organized a prize contest for a base plan for the development of Greater Berlin. This involved not just public transport, but also projections of land use, population development, and spatial planning. The results of this *Wettbewerb Groß-Berlin* came to be shown to the public at the University of Fine Arts in Charlottenburg from May to June 1910, thereby creating a great amount of

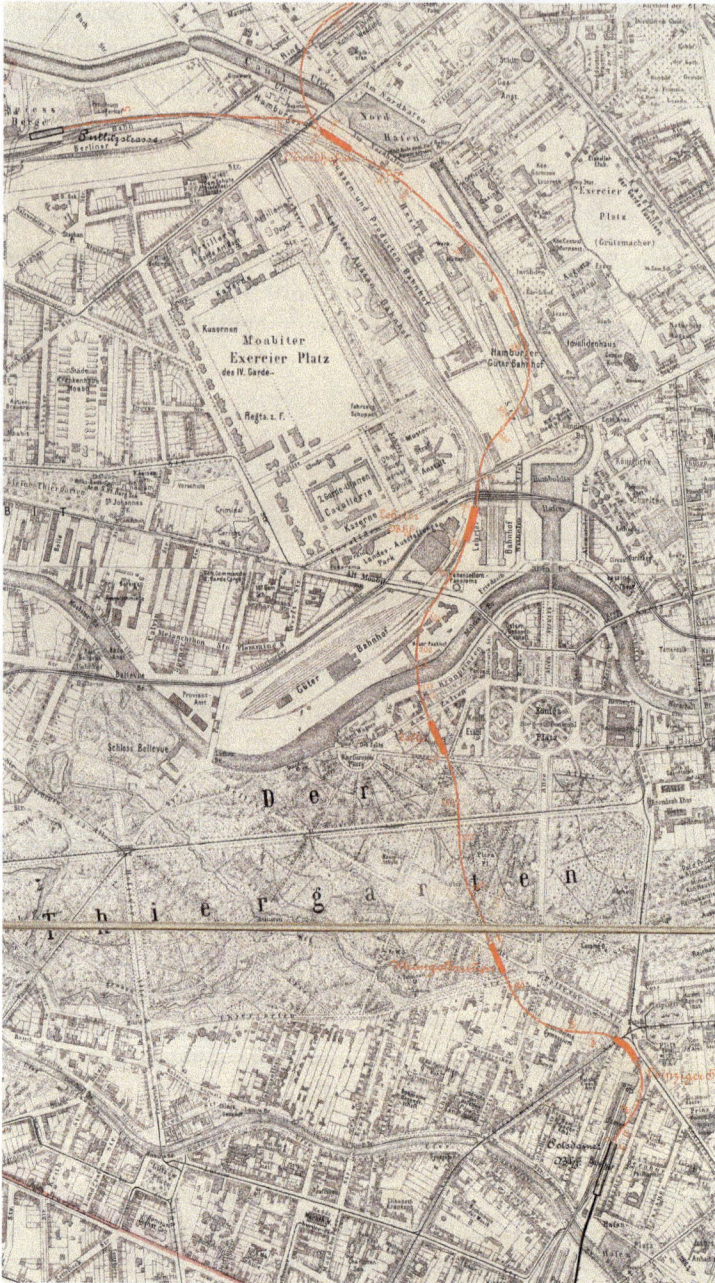

Fig. 11.2: Gustav Schimpff, proposed route (left; detail) of a north-south line and design for the station "Zelte" (right) on the elevated railway through Berlin, 1897.

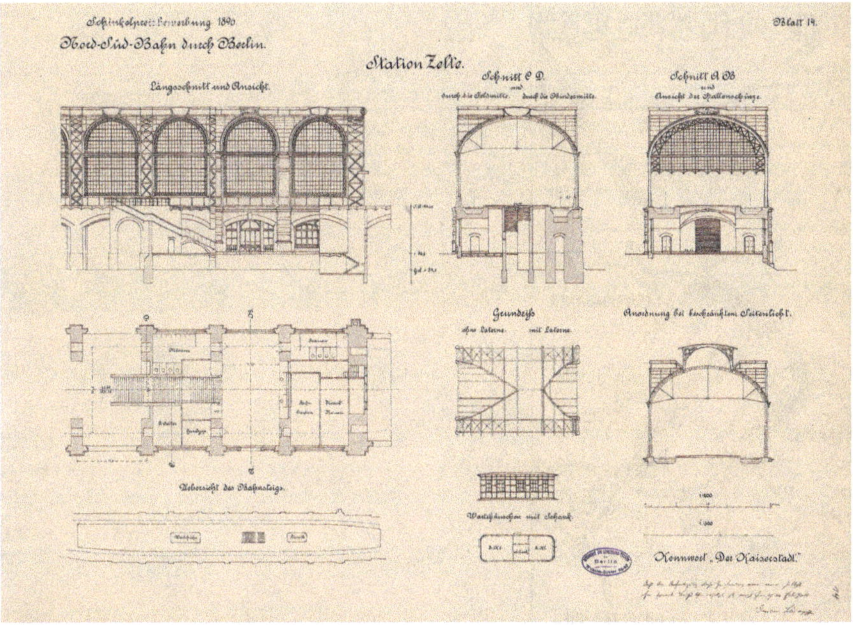

Fig. 11.2 (continued)

discussion and, predictably, controversy. The first and second prizes were shared between Hermann Jansens (who gained 17 votes from the jury) and the duo of Josef Brix and Felix Genzmer (who received 12). Transit was not all that central to Jansens' concept, but he did forward a plan for a grouping together of Potsdamer and Anhalter in the south of the city, and a tunnel connection to a new station on the site of the Lehrter Bahnhof. That solution drew the ire of Gustav Schimpff, who proclaimed Jansens' work unworthy of serious discussion.

Brix and Genzmer, whose plan had been developed together with the Berlin Elevated Railway Company (Hochbahngesellschaft), did not address the issue of transport themselves; rather, they had left it to several engineers from the Hochbahngesellschaft. In addition to a general rationalization, they proposed several practical solutions, including the construction of an underground connection between Pankow and Tempelhof via the Ringbahn, Lehter Bahnhof, and Potsdamer Bahnhof. Simultaneously, they suggested an S-Bahn tunnel leading from the Potsdamer Bahnhof to Gesundbrunnen via Friedrichstraße, an idea that had already been proposed and would actually be built in the 1930s.

Some other contributions went further. Third-prize winners Eberhard, Möhring, and Petersen wanted to move the entire Potsdamer Bahnhof with all its tracks underground, with a glass above-ground entrance building (remarkably

Fig. 11.3: Gustav Kemmann, Paul Wittig, Emil Pavel, Johannes Bousset, Emil Bandekow, and Heinrich Schmidt, proposed route for the railways in central Berlin Die preisgekrönten.

similar to today's situation), although it would still have been a terminus, which would have freed up space for urban development in the huge railway yard behind the station. However, the most profound reorganisation was conceived by fourth prize winners Havestadt, Contag and Schmitz. They proposed the centralization of services in a main south and north station, the latter on the Stadtbahn,

Fig. 11.4: Havestadt, Contag & Schmitz, New Berlin (section) around the Northern Central Station (right). Submission for the Wettbewerb Groß-Berlin 1910. Neu-Berlin am Nordzentralbahnhof.

which would serve as termini for long-distance trains. Local trains, meanwhile, would be led through an underground tunnel linking both.

This idea was further developed by city planner Martin Mächler. Originally intended for the *Groß-Berlin* competition but never sent in, his paper repeated several ideas that had been floated earlier, such as the north-south railway, but it also added new elements. The most notable one was a monumental avenue for road traffic running north to south, crossing the Charlottenburger Chaussee (the continuation of Unter den Linden) just before the Brandenburg gate. Like Schmitz et al., Mächler conceived two large railway stations on either end of this new axis: one in the north, in the location of the Lehrter Bahnhof along the Stadtbahn; the other in the south. The Anhalter Bahnhof was to become a large indoor swimming pool, whereas the Potsdamer Bahnhof would be removed altogether and replaced by a smaller local stop along an underground railway connecting the two major stations.

A Capital for a Thousand Years

Mächler's axis made a return in the plan that has become the most famous of all schemes to reorganize Berlin, concocted by Adolf Hitler and Albert Speer to create Germany's new capital, the *Welthauptstadt* (world capital) *Germania*. Far outdoing other Nazi construction efforts such as Hitler's Reich Chancellery or Goering's Air Ministry in scale, it was nothing if not ambitious. Allegedly conceived by Adolf Hitler as early as the 1920s the Germania plan was centered around a central avenue running north-south and connecting a new southern station and a smaller but still enormous Northern Station.

Fig. 11.5: Model of the planned Welthauptstadt Germania, view from the southern railway station over the triumphal arch to the Great Hall (north-south axis). Note the Anhalter Bahnhof near the right edge of the image, about two-thirds up from the bottom. Part of the exhibition Mythos "Germania" and "Tempelstadt" Nuremberg at the Nuremberg Documentation Center Nazi Party Rally Grounds (2011).

Hitler and Speer envisioned nothing less than a redesign of the capital, but in terms of railway infrastructure they based themselves mostly on the Mächler Plan. The main change was that both the north and south station were located further away, the first one on top of the northern Ringbahn, the second also on the Ringbahn, near Tempelhof Airport in the south. This emphasized what mainly distinguished the Nazis' plans from earlier ones: scale. The center of the avenue was to be dominated by the Hall of the People, a huge domed construction designed to

Fig. 11.6: Impression of the vast interior of the planned South Station.

contain over a quarter of a million people for Nazi party rallies. Going south, traffic was going to drive down a 60-meter wide avenue to pass under a triumphal arch that would have allowed Paris's Arc de Triomphe to pass under it as well.

Everything was set to be gargantuan and then some, and the stations were no exception. The South Station was planned as the main rail hub, with ten platforms covered by a 70-meter high train shed measuring 400 meters in length. The North Station was slightly less enormous, with a similarly sized shed being a mere 54 meters in height. Included in the plans were West and East stations, modest by comparison but still much larger than the ones they replaced.

Some preparatory work was done for the construction of Germania, such as widening the Charlottenburger Chaussee (today's Straße des 17. Juni), moving the Victory Column (Siegessäule) to its present location, and tearing down the Alsenviertel, the embassy and government quarter in front of the Lehrter Bahnhof. However, the outbreak of open war in 1940 prevented further execution, and by the end of 1943 the team working on the project had been disbanded and construction halted. Like most things connected to the Nazi regime, their projected city retained its popular fascination, and Germania has seen (usually somewhat inaccurate) depictions in multiple films and television productions, most recently in Netflix's *The Man in the High Castle*, based on the eponymous book by Philip K. Dick.

After Germania

After Germany's defeat in the Second World War, the extensive wartime damage had also simplified the chaotic nature of the system. The Potsdamer Bahnhof was taken out entirely, while traffic to the remaining termini diminished significantly. The city's pre-war population of almost four and a half million had been reduced to under three million in late 1945, which meant that for the first time in its history, there was no longer the constant pressure of growth in passenger numbers. The GDR prohibited its citizens to travel to West Berlin in 1952 and made it altogether impossible to do so by building the Wall in 1961, which also meant that most of the old termini got cut off from the tracks that served them. The Stadtbahn now truly became the city's railway artery as, one by one, the other stations shut down.

The era of planning for an even greater Berlin was over as political circumstances fluctuated initially before settling down to their uncomfortable but at least relatively stable status quo during the 1960s. Considerable stretches of tracks had been discarded or transported to the Soviet Union as war reparations; moreover, car-centric city planning became the norm, particularly in West Berlin. As highways were being constructed as its main transportation infrastructure, few city planners cared much for what was regarded to be an antiquated form of transport. In addition, further development of the railways proved problematic; the East Germans ran the S-Bahn (up to 1984) and apart from the inter-zone trains no long-distance traffic was allowed outside the walled-in city. Although the subways were extended on both sides of the Wall, little love was given to the regular railways.

In the east of the city, the situation was somewhat different. The necessity to avoid West Berlin led to the completion of an outer railway ring, the Berliner Außenring, around the whole city in 1961. People could now travel from East Berlin to places in the GDR west of West Berlin, even if often by a rather tortuous route. And while East Berlin was by no means immune to the automobilophilia of the 1960s, practical circumstances forced it to place more faith in railway transportation.

A New Hope

Of course, the situation again changed dramatically on 9 November 1989, when the Berlin Wall ceased to be the insurmountable obstacle it had been for 18 long years in a single evening. In quick succession, the Democratic Republic and Federal Republic were merged, the wall removed, and Berlin declared the capital of the newly unified Germany. With it came a new stage in planning for Berlin's future transit needs.

At the time of German re-unification in 1990, the capital's rail systems were in dire need of maintenance and renovation. Parts of the system had still not been restored after war damage, and cold war measures needed to be rolled back to simply bring it back to working order. It took about a decade to fully restore the Ringbahn of 1871, but in 1998 passengers could make the full loop around the center again for the first time since the 1950s. The underground "Ghost stations," where trains were not allowed to stop crossing beneath East Berlin, were also reinstated, and little by little the city's urban rail was getting back to pre-war normality.

Simultaneously it became glaringly obvious that facilities for long-distance connections required a more drastic approach. The main stations for both halves of the separated city – Zoologischer Garten in the West and Hauptbahnhof (formerly Ostbahnhof) in the East – were entirely insufficient to service a city of over three million people. After the decision was taken to make Berlin the German capital once again, arguments of political prestige entered the arena: none of the existing stations was in any way properly representative for the capital of a newly unified Germany, Europe's most populous and economically most significant nation. Moreover, it was – correctly – assumed that the number of people visiting Berlin would increase significantly.

A fundamental problem was that the capacity of the Stadtbahn, the backbone of the railways traversing the city center since 1882, was inherently limited because of its four tracks, and could not easily be expanded without extensive demolition. While this problem could be alleviated by using other railways in and around Berlin, that approach might easily result in the same kind of fragmentation that so many had tried to resolve before the Second World War; for the same reason, a reconstruction of the old system was never seriously considered. After extensive discussions in the early 1990s, a consensus had formed in favor of a single large public transport hub. The obvious location was the one that had been so prevalent in previous plans, that of the now-demolished Lehrter Bahnhof. Another possibility in a more central location would be Friedrichstraße but there space was much more of an issue, as the Friedrichstraße was already undergoing frantic development.

In 1992, the railway authorities and the Berlin City Senate therefore decided on the site of the Lehrter Bahnhof as the location for a new Hauptbahnhof, something the city had never really possessed. The central element in the plans for the new main station became the so-called "mushroom concept" (Pilzkonzept), an adaptation of the idea that had done the rounds at least since Schimpff's plan of 1897. The name refers to the shape of the lines, with the northern half of the Ringbahn/Stadtbahn combination symbolizing the head of the mushroom, whereas the stem is represented by the southern-bound line towards Südkreuz and on to Leipzig and Dresden.

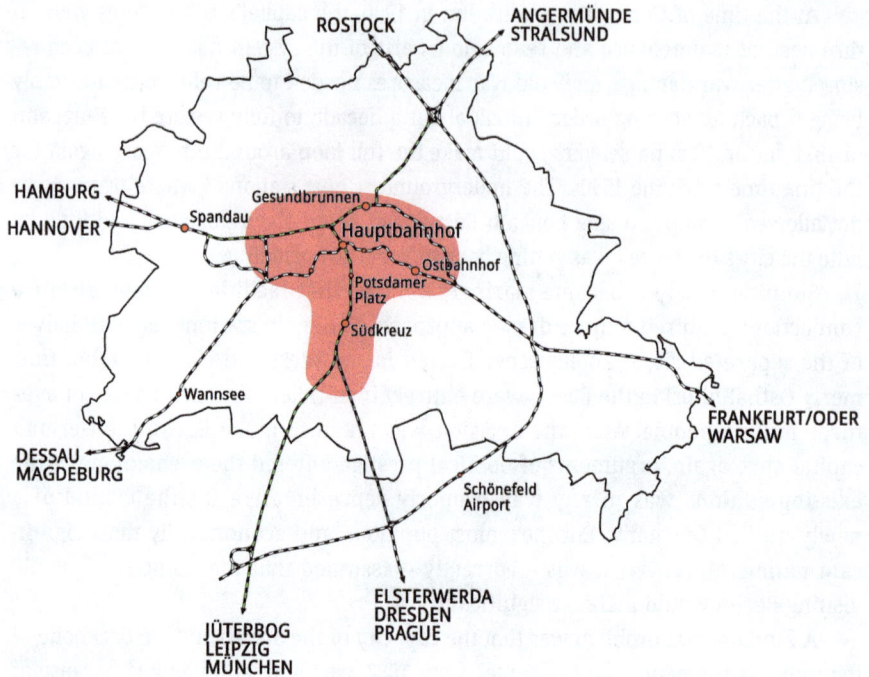

Fig. 11.7: Berlin's "Mushroom" concept.

The prize contest for the new station was won in early 1993 by the architectural firm of Gerkan, Marg, and Partner, with Meinhard von Gerkan serving as the chief architect of a station with crossing lines – known in Germany as a Turm-bahnhof or "Tower Station". The upper levels were to be integrated into the Stadt-bahn, whereas the lower levels were to use a new tunnel, with platforms in more or less the same place as the old Lehrter Bahnhof had once occupied, albeit several meters down. A three-and-a-half-kilometer-tunnel, already an element of the Jansens Plan, was to connect to Potsdamer Platz (where a new station was built close to the location where the Potsdamer Bahnhof had once stood) and from there on southward.

The main victim of the new Hauptbahnhof was the Lehrter Stadtbahnhof. It was one of three stations still left in its original state from the construction of the Stadt-bahn in 1882. Although protected, it was nevertheless demolished to make way for the new station in 2002. Whether this is seen as a senseless act of cultural vandalism or a necessary step in the rationalization of transport in the new capital depends on one's viewpoint, but the casualness with which the decision was taken by Germany's national rail company Deutsche Bahn creates uncomfortable associations with the demise of previous railway monuments such as the Anhalter and Görlitzer stations.

Architect and railway company did not always see eye to eye. Gerkan's dismissive attitude towards retail space, which he regarded as a dilution of his design, was opposed by Deutsche Bahn, who considered shops as essential for the exploitation of the station. A more lasting point of contention was the length of the upper canopy. Originally planned to be over four hundred meters long, its length was cut back to 321 meters, much to the chagrin of Gerkan. The decision was explained with the wish to have the station in operation during the 2006 FIFA World Cup, however, the growing cost of the project appears to have been a more important factor; Gerkan sued and won, but until today there are no plans to extend the canopy to its projected length. The 800 million euro budget would eventually end up at around 1,2 billion euros; the projected opening date in 2003 was not met either, with the station finally entering service in 2006 to widespread praise and, likely, relief that finally and after 150 years of planning, Berlin had finally received a main station worthy for a capital.

Berlin Hauptbahnhof, a Critique and Epilogue

What is the end result? First of all, most will likely agree that, whatever one thinks of its execution, Berlin's present rail infrastructure is the best it has ever

Fig. 11.8: Berlin Hauptbahnhof in 2024.

had. The choice for the "Mushroom" has proven essential for guaranteeing adequate rail transport for the capital, at least for now; even today the Stadtbahn can barely manage the intense traffic, and the North-South corridor is also used heavily. In hindsight, and over 15 years after its completion, building Hauptbahnhof in its present location seems to have been a good idea.

However, that is not to say that some issues do not remain. First of all, Hauptbahnhof is hardly a truly *central* station. One of the issues is the same that plagued its predecessor, the Lehrter Bahnof; it is located in a place that only merits a visit when one needs to be at the station, and at least at present there is not much else around. If we are fair, that is perhaps not really a major problem in a decentralized city such as Berlin, where no station could ever be truly central. The quality of the station's experience therefore largely depends on the quality of its connections.

This was a real issue at the time of Hauptbahnhof's opening, because apart from train links and the S-Bahn lines on the Stadtbahn, there was no connection to other forms of public transport other than buses. Since then, a lot has improved, most importantly with the connection to the U-Bahn. Although ludicrously short (and late) when it opened in 2009, the U55 at least established a connection to the north-south lines of the S-Bahn. Now, after the creation of a full connection to the U5 line in late 2020, the station possesses a truly excellent link to the old center of the city as well as the rest of the transit system. In 2014, a

Fig. 11.9: View of the lower platform level of Berlin Hauptbahnhof.

tram connection came into service, and despite delays it looks like a second S-Bahn connection will be opened sometime in the late 2020s.

But while the problem of connections is being gradually resolved, there are still some fundamental issues that haunt the station. One is that classical issue of all of Berlin's railway infrastructure: growth. Passenger numbers have risen by around 30% in the last decade, and with around 330,000 post-Covid travelers converging on the station every day, it is getting quite cramped. Built atop the Stadtbahn of 1882, the upstairs platforms are precariously narrow for coping with the stream of passengers; this is where all the east-west connections arrive and depart, and the amount of people wanting to board trains, along with their luggage, stretches the space to capacity.

Gerkan tried to do what he could, creating additional space inside the curve. However, this meant tightening it, which created other problems, such as increased noise from the train wheels. Their screech often drowns out announcements, but as those are mostly inaudible anyhow because of the shed's fittingly cathedral-like acoustics, it is perhaps not a great loss. Also, the Deutsche Bahn's penny-pinching decision to shorten the platform canopies means that passengers often await their train in the open air, weathering everything the elements can throw at them, an inadvertent callback to the circumstances at Berlin's first station in 1838.

Fig. 11.10: Inside Berlin Hauptbahnhof.

There is, however, one thing that we can unequivocally blame Gerkan for: Hauptbahnhof's design – or rather, its floor plan. The issue here is not an aesthetic one, but rather one of orientation or navigation. Because of the multiple levels, the multidirectional escalators and the plethora of impressions, newcomers and even returning travelers to Hauptbahnhof are easily confused and lose sight of which side of the station is which. Good spatial design is essential in places where lots of people try to go from one place to the next, and Berlin Hauptbahnhof drops the ball in this area.

People may point toward the size of the station, and while it is massive there are examples of similarly sized stations that do not have this problem. Utrecht Centraal in the Netherlands, an even larger hub with a multitude of transport options, is surprisingly straightforward to navigate. On the other hand, Gesundbrunnen, just north of Hauptbahnhof, is much smaller than the main station but somehow manages to be even more confusing. Good spatial design does not care about size.

What it does care about is transparency. A good station is designed in such a way that it immediately and intuitively becomes obvious what is where. Each of Berlin Hauptbahnhof's six levels offers either no view of the others, or a very limited one. Consequently, it is difficult to make out where you are, let alone where you must go. Commercialization plays its part here, too; there is no single floor with shops: they are strewn everywhere, indiscriminately it seems, so they cannot act as a point of orientation. Wayfaring is chaotic, inconsistent, and worst of all inconspicuous.

Nonetheless, it can't be denied that Hauptbahnhof has also been a success, and without it Berlin's train networks would be in even greater trouble than they are with it. It has become Germany's fourth-busiest railway station after Hamburg, Frankfurt, and Munich, and an essential part of the city's and the country's rail system. In a way, it is satisfying to see that Berlin Hauptbahnhof is being confronted with some of the city's traditional troubles; as the population of the German capital is again on the rise, and as post-Covid travel and tourism increases as well, we may not be so far from a point at which new plans need to be made, and the form or even existence of Hauptbahnhof may once again become a topic of debate.

Because if there's one thing Berlin will never suffer, it is complacency.

12 Further Reading

A Note on Statistics

The sources for statistical information specific to Berlin's railways are surprisingly limited. The figures used in this book are based primarily on the 1896 report on Berlin's railways (*Berlin und seine Eisenbahnen 1846 – 1896*. Im Auftrage des Verein Deutscher Eisenbahnverwaltungen, Verband der Preussischen Eisenbahnen, Königlich Preussischer Minister der Öffentlichen Arbeiten. Two volumes. Berlin & Heidelberg: Springer, 1869) and Louis Jänicke's article from 1924 (Jänicke, Louis. "Der Personenverkehr in Berlin." *Verkehrstechnische Woche* 18 (1924): 395–401). Fremdling et al.'s volume on statistics of German railways (Fremdling, Rainer, Ruth Federspiel, and Andreas Kunz, eds. *Statistik der Eisenbahnen in Deutschland, 1835–1989*. St. Katharinen: Scripta Mercatura, 1995) also offers a wealth of data.

Fig. 12.1: Hamburger Bahnhof in Berlin Johann Friedrich Stiehm, 1868–1870, stereograph.

Anyone interested in the nineteenth century railway (and transport) experience should read Wolfgang Schivelbusch's *The Railway Journey* (German edition 1981, updated English edition 2014), chapters 11–13. The most extensive treatment of German station architecture in the nineteenth century is probably still Ulrich Krings' *Deutsche Großstadt-Bahnhöfe des Historismus* (1981 as a PhD dissertation, with a commercial edition in 1985) and I warmly recommend it to anyone interested in the subject. For more detailed information, I refer you to the works of Peter Bley and particularly the doyen of Berlin's public transport history, the late Alfred Gottwaldt.

There are a few detailed and excellent descriptions of particular stations. I certainly wish to mention Helmut Maier's extensive history of the Anhalter Bahnhof, Laurenz Demps' book on the Schlesischer Bahnhof, Lothar Uebel's history of the Küstriner Bahnhof, and Holger Steinle's book about the Hamburger Bahnhof. All of these titles are somewhat older (and therefore only available in used copies) and only available in German; part of my purpose here was to make their contents

https://doi.org/10.1515/9783111381879-012

available to English speakers. For those more interested in the history of rolling stock, there is a hoard of material available, but a good starting point is Winfried Reinhardt's *Geschichte des öffentlichen Personenverkehrs von den Anfängen bis 2014.*

00 The Best Books About Railway Stations in Berlin

Demps, Laurenz. *Der Schlesische Bahnhof in Berlin. Ein Kapitel preußischer Eisenbahn-Geschichte.* Berlin: TransPress, 1991. An exhaustive history of today's Ostbahnhof, from possibly the best living author on the subject.

Gottwaldt, Alfred B. *Das Große Berliner Eisenbahn-Album.* Stuttgart: TransPress, 2010. A richly illustrated overview of the history of Berlin's railways by arguably the foremost expert in the field.

Krings, Ulrich. *Deutsche Großstadt-Bahnhöfe des Historismus.* München: Prestel, 1985. Commercial publication of Kring's PhD overview of historicist station architecture. Covers all of Germany, but with a heavy emphasis on Berlin. Fairly technical, but thorough and solid.

Maier, Helmut. *Berlin Anhalter Bahnhof.* Berlin: Verlag Ästhetik und Kommunikation, 1984. The standard work on the Anhalter's history, with a lot of information about Berlin's other stations.

Steinle, Holger. *Ein Bahnhof auf dem Abstellgleis. Der Ehemalige Hamburger Bahnhof in Berlin und seine Geschichte.* Berlin: Silberstreif, 1983. A thorough history of the station, written in a time in which the wall was still standing but the Hamburger Bahnhof almost wasn't.

Uebel, Lothar. *Eisenbahner, Artisten und Zeitungsmacher. Zur Geschichte des ehemaligen Küstriner Bahnhofs.* Berlin: Grundstücksgesellschaft Franz-Mehring-Platz, 2011. A fun story about probably the most extraordinary station history in the city.

0 General Literature

Berger, Manfred. *Historische Bahnhofsbauten Sachsens, Preussens, Mecklenburgs und Thüringens.* Berlin: Transpress, 1981.

Brauchitsch, Boris von. *Unter Dampf: historische Fotografien von Berliner Fern- und Regionalbahnhöfen.* Berlin: Braus, 2018.

Deutsche Bahn AG. *Planet Eisenbahn: Bilder und Geschichten aus 175 Jahren.* Köln/Wien: Böhlau Verlag, 2010.

Gottwaldt, Alfred B. *Eisenbahn-Brennpunkt Berlin. Die deutsche Reichsbahn 1920–1939.* Stuttgart: Franckh, 1976.

Gottwaldt, Alfred B. Berliner Fernbahnhöfe. Erinnerungen an ihre große Zeit. Düsseldorf: Alba, 1982.

Gottwaldt, Alfred B. *Das große Berliner Eisenbahn-Album.* Stuttgart: TransPress, 2010.

Knipping, Andreas. *175 Jahre Eisenbahn in Deutschland. Die Illustrierte Chronik.* München: GeraMond, 2010.

Neumann, Peter. *Berlins Bahnhöfe. Gestern, heute, morgen.* Berlin: Jaron Verlag, 2004.

Reinhardt, Winfried. *Geschichte des öffentlichen Personenverkehrs von den Anfängen bis 2014. Mobilität in Deutschland mit Eisenbahn, U-Bahn, Straßenbahn und Bus.* Wiesbaden: Springer Vieweg, 2015.

Sauer, Mark. In geplanten Bahnen: Eisenbahnanlagen als Kulturlandschaftselemente in Deutschland von 1848 bis 1998. Bonn: Universität Bonn, 2000.

Schivelbusch, Wolfgang. *The Railway Journey.* Berkeley & Los Angeles: University of California Press, 2014.

Winkler, Dirk. Eisenbahnmetropole Berlin 1894 bis 1934. Berlin: EK-Verlag, 2015.

1 Introduction

Bley, Peter. *150 Jahre Eisenbahn Berlin-Potsdam. Aus der Geschichte der ältesten Eisenbahn in Berlin und Preussen.* Düsseldorf: Alba, 1988.
Bley, Peter. *150 Jahre Eisenbahn Berlin – Frankfurt/Oder.* Düsseldorf: Alba Publikation, 1992.
Bley, Peter. *150 Jahre Eisenbahn Berlin – Hamburg.* Berlin: Alba, 1996.
Bley, Peter. 125 *Jahre Berlin-Dresdener Eisenbahn. Berlin-Zossen-Elsterwerda-Dresden.* Düsseldorf: Alba, 1999.
Bley, Peter. *Berliner Nordbahn: 125 Jahre Eisenbahn Berlin-Neustrelitz-Stralsund.* Berlin: B. Neddermeyer, 2002.
Demps, Laurenz, "Vom Frankfurter Bahnhof zum Hauptbahnhof. Aus der Geschichte des Berliner Ostbahnhofs." *Modell-Eisenbahn* 9, no. 87 (1987).
Gall, Lothar. "Eisenbahn in Deutschland: von den Anfangen bis zum Ersten Weltkrieg." In *Die Eisenbahn in Deutschland. von den Anfangen bis zur Gegenwart,* edited by Lothar Gall and Manfred Pohl, 13–74. München: C.H. Beck, 1999.
Jänicke, Louis. "Der Personenverkehr in Berlin." *Verkehrstechnische Woche* 18 (1924): 395–401.
Krause, Falko. *Die Stadtbahn in Berlin. Planung, Bau, Auswirkungen.* Hamburg: Diplomica, 2014.
Maier, Helmut. *Berlin Anhalter Bahnhof.* Berlin: Verlag Ästhetik und Kommunikation, 1984.
Nilsen, Micheline. *Railways and the Western European Capitals. Stories of Implantation in London, Paris, Berlin, and Brussels.* New York: Palgrave Macmillan, 2008.
Richards, Jeffrey, and John M. MacKenzie. *The Railway Station: A Social History.* Oxford & New York: Oxford University Press, 1986.
Steinle, Holger. *Ein Bahnhof auf dem Abstellgleis. Der Ehemalige Hamburger Bahnhof in Berlin und seine Geschichte.* Berlin: Silberstreif, 1983.
Uebel, Lothar. *Eisenbahner, Artisten und Zeitungsmacher. Zur Geschichte des ehemaligen Küstriner Bahnhofs.* Berlin: Grundstücksgesellschaft Franz-Mehring-Platz, 2011.
Winkler, Dirk. *Die Eisenbahn in Berlin,* Eisenbahn-Kurier Special 133 (2019).
Zschocke, Helmut. *Die Berliner Akzisemauer: die vorletzte Mauer der Stadt.* Berlin: Berlin Story Verlag, 2007.
Zschocke, Helmut. Die erste Berliner Ringbahn: Über die Königliche Bahnhofsverbindungsbahn zu Berlin. Berlin: Verlag Bernd Neddermeyer, 2009.

2 Potsdamer Bahnhof

"Wettbewerb für Vorentwürfe zur Neugestaltung des Vorplatzes am Potsdamer Hauptbahnhof in Berlin." *Zentralblatt der Bauverwaltung* 39 (1919): 585–89.
Bley, Peter. *150 Jahre Eisenbahn Berlin-Potsdam. Aus der Geschichte der ältesten Eisenbahn in Berlin und Preussen.* Düsseldorf: Alba, 1988.
Bock, Hans. "Entstehung und Geschichte der Eisenbahn in Berlin (1838–1961)." *Jahrbuch für Eisenbahngeschichte* 11 (1979): 5–48.
Handke, Peter. *Die Eisenbahn Berlin-Potsdam. Die Wannseebahn.* Berlin: Marion Hildebrand Verlag, 1988.
Krings, Ulrich. *Deutsche Großstadt-Bahnhöfe des Historismus.* München: Prestel, 1985.
Winkler, Dirk. "Potsdamer Bahnhof – der Dreigeteilte." *Die Eisenbahn in Berlin, Eisenbahn-Kurier Special* 133 (2019): 44–53.

3 Anhalter Bahnhof

Berger, Manfred. *Historische Bahnhofsbauten Sachsens, Preussens, Mecklenburgs und Thüringens*. Berlin: Transpress, 1981.
Berggruen, Heinz. *Kleine Abschiede: 1935–1937: Berlin, Kopenhagen, Kalifornien*. Berlin: Transit, 2004.
Bley, Peter. *125 Jahre Berlin-Dresdener Eisenbahn. Berlin-Zossen-Elsterwerda-Dresden*. Düsseldorf: Alba, 1999.
Boberg, Jochen. *Der Anhalter. Geschichte und Geschichten um den größten Bahnhof Berlins*. Berlin: Museumspädagogischer Dienst Berlin, 1983.
Demps, Laurenz. *Der Schlesische Bahnhof in Berlin. Ein Kapitel preußischer Eisenbahn-Geschichte*. Berlin: TranzPress, 1991.
Eiselen, Fritz. "Die Lösung der Verkehrsfragen im Wettbewerb Groß-Berlin." *Deutsche Bauzeitung 1910*, no. 50–59 (1910): 385–392, 401.
Fritsch, Karl. "Das neue Empfangs-Gebäude der Berlin-Anhaltischen Eisenbahn in Berlin." *Deutsche Bauzeitung* 13 (1879): 11–14, 23–25, 41.
Gympel, Jan. "Wie der Postbahnhof zum Dresdener Bahnhof wurde Die DB AG entdeckte einen neuen alten Bahnhof – und hat sich damit gründlich blamiert [signalarchiv.de]." *Signal 1997*, no. 7 (1997): 9–11.
Hofmann, Albert. "Groß-Berlin, sein Verhältnis zur modernen Großstadtbewegung und der Wettbewerb zur Erlangung eines Grundplanes für die städtebauliche Entwicklung Berlins und seiner Vororte im zwanzigsten Jahrhundert." *Deutsche Bauzeitung* 44 (1910): 169–181, 197, 213, 233, 261, 281, 311, 325.
Knothe, Rainer. *Anhalter Bahnhof: Entwicklung und Betrieb; Zeugen und Zeugnisse aus über 100 Jahren*. Berlin: Verlag Ästhetik und Kommunikation, 1987.
Krings, Ulrich. *Bahnhofsarchitektur. Deutsche Großstadtbahnhöfe des Historismus*. München: Prestel, 1985.
Maier, Helmut. *Berlin Anhalter Bahnhof*. Berlin: Verlag Ästhetik und Kommunikation, 1984.

4 Stettiner Bahnhof

Berger, Manfred. *Historische Bahnhofsbauten Sachsens, Preussens, Mecklenburgs und Thüringens*. Berlin: Transpress, 1981.
Berlin und seine Eisenbahnen 1846 – 1896. Im Auftrage des Verein Deutscher Eisenbahnverwaltungen, Verband der Preussischen Eisenbahnen, Königlich Preussischer Minister der Oeffentlichen Arbeiten. Two volumes. Berlin & Heidelberg: Springer, 1896.
Bley, Peter. *Berliner Nordbahn: 125 Jahre Eisenbahn Berlin-Neustrelitz-Stralsund*. Berlin: B. Neddermeyer, 2002.
Bock, Hans. "Entstehung und Geschichte der Eisenbahn in Berlin (1838–1961)." *Jahrbuch für Eisenbahngeschichte* 11 (1979): 5–48.
Brauchitsch, Boris von. *Unter Dampf: historische Fotografien von Berliner Fern- und Regionalbahnhöfen*. Berlin: Braus, 2018.
Braun, Michael. *Nordsüd-S-Bahn Berlin. 75 Jahre Eisenbahn im Untergrund*. Berlin: Gesellschaft für Verkehrspolitik und Eisenbahnwesen, 2008.

Cornelius, Carl. "Um- und Erweiterungsbau des Empfangsgebäudes auf dem Stettiner Bahnhof in Berlin." *Zeitschrift für Bauwesen* 54 (1904): 213–24.

Demps, Laurenz. *Der Schlesische Bahnhof in Berlin. Ein Kapitel preußischer Eisenbahn-Geschichte.* Berlin: TranzPress, 1991.

Gröninck, E. *Berlin und seinen zukünftigen Central-Bahnhofs- und Central-Hafen-Anlagen.* Berlin: Polytechnische Buchhandlung A. Seydel, 1901.

Hallfahrt, Hans-Günter. "Berliner Eisenbahnen und ihre Bahnhöfe von den Anfängen bis 1870." *ICOMOS – Hefte des Deutschen Nationalkomitees* 4 (1992): 50–52.

Koll and Helm. "Der Verkehr in Gross-Berlin." *Verkehrstechnische Woche* 5, no. 11, 14, 21, 26, 28, 30 (1911): 260–266, 513–520, 345–347, 693–702, 744–750.

Kuhlmann, Bernd. *Die Berliner Bahnhöfe.* München: Geramond, 2010.

Preuß, Erich. "Max Palme reist nach Stettin." *Bahn-Special* 1, no. 95 (1995): 38–41.

Regling, Horst, Dieter Grusenick, and Erich Morlok. *Die Berlin-Stettiner Eisenbahn.* Stuttgart: Transpress, 1996.

Roos, Harald. "Berlins Fern- und Nahverkehr. Vorschlag zu einer grundlegenden Neuordnung." *Das neue Berlin* 1, no. 7 (1929): 142–43.

Stürickow, Regina. *Mörderische Metropole Berlin. Authentische Fälle 1914–1933.* Leipzig: Militzke, 2004.

Winkler, Dirk. "Der Ferienbahnhof." *Die Eisenbahn in Berlin.* Eisenbahn-Kurier Special 133 (2019): 30–35.

Worm, Hardy. *Rund um den Alexanderplatz.* Berlin: Aufbau Verlag, 1981.

5 Schlesischer Bahnhof

"Verein für Eisenbahnkunde zu Berlin. Hr. Römer gab eine durch vorgelegte Zeichnungen erläuterte Beschreibung des im Bau begriffenen neuen Stationsgebäudes auf dem hiesigen Bahnhofe der Niederschlesisch-Märkischen Eisenbahn." *Deutsche Bauzeitung* 2 (1868): 111–12.

Bley, Peter. *150 Jahre Eisenbahn Berlin – Frankfurt/Oder.* Düsseldorf: Alba Publikation, 1992.

Bosetzky, Horst. *Die Bestie vom Schlesischen Bahnhof.* Berlin: Jaron-Verlag, 2004.

Demps, Laurenz. *Der Schlesische Bahnhof in Berlin. Ein Kapitel preußischer Eisenbahn-Geschichte.* Berlin: TransPress, 1991.

Erbkam, G. "Schlesischer Bahnhof, Berlin." *Atlas zur Zeitschrift für Bauwesen* 34 (1884):

Fink, Georg. *Menschen am Schlesischen Bahnhof.* Berlin: Bruno Cassirer, 1930.

Moldt, Dirk. "Abgehängt trotz seiner sechs Namen." Friedrichshainer ZeitZeiger. 2017. Accessed 14 December 2022. https://fhzz.de/ostbahnhof/view-all/.

Nolte, Paul. *Die Vergnügungskultur der Großstadt: Orte – Inszenierungen – Netzwerke (1880–1930).* Köln/Wien: Bohlau Verlag, 2015.

Römer, Eduard. "Umbau des Bahnhofes der Niederschlesisch-Märkischen Eisenbahn zu Berlin." *Zeitschrift für Bauwesen* 20 (1870): 151–72.

Schaulinsky, Gernot. "1880–1920: Rund Um Den Wilhelminischen Ring." Ringbahn.com. 2010. Accessed 27 June 2022. http://www.ringbahn.com/rueckblick-1880-1920.html.

6 The Stadtbahn

"100 Jahre Berliner Stadtbahn." *Verkehrsgeschichtliche Blätter* 9, no. 7 (1982): 158–89.

Berggruen, Heinz. *Kleine Abschiede: 1935–1937: Berlin, Kopenhagen, Kalifornien*. Berlin: Transit, 2004.

Dominik, Emil. *Quer durch und ringsum Berlin. Eine Fahrt auf der Berliner Stadt- und Ringbahn*. Berlin: Gebrüder Patel, 1883.

Koschinsky, Konrad. "Die Stadtbahn – 12 Kilometer mitten durch die Mitte." *Bahn-Special* 1, no. 95 (1995): 18–22.

Krause, Falko. *Die Stadtbahn in Berlin. Planung, Bau, Auswirkungen*. Hamburg: Diplomica, 2014.

Krings, Ulrich. "Die Stammstrecke der Berliner Stadtbahn – eine technische, urbanistische und architektonische Leistung des 19. Jahrhunderts." In *Die Berliner S-Bahn. Gesellschaftsgeschichte Eines Industriellen Verkehrsmittels*, edited by Knut Hickethier, 73–81. Berlin: Verslag Ästhetik und Kommunikation, 1982.

Rudolf, H. "Auf der Berliner Stadtbahn." *Westermann's Illustrierte Deutsche Monatshefte* 309 (1882): 368–80.

Schaulinsky, Gernot. "1880–1920: Rund Um Den Wilhelminischen Ring." Ringbahn.com. 2010. Accessed 27 June 2022. http://www.ringbahn.com/rueckblick-1880-1920.html.

Zschocke, Helmut. Die erste Berliner Ringbahn: Über die Königliche Bahnhofsverbindungsbahn zu Berlin. Berlin: Verlag Bernd Neddermeyer;, 2009.

7 Hamburger Bahnhof

Bley, Peter. *150 Jahre Eisenbahn Berlin – Hamburg*. Berlin: Alba, 1996.

Brühl, Christine. *Der Hamburger Bahnhof. Der historische Ort*, vol. 53. Berlin: Kai Homilius, 2003.

Hallfahrt, Hans-Günter. "Berlin – Eisenbahn und Stadtentwicklung." *ICOMOS – Hefte des Deutschen Nationalkomitees* 9 (1993): 48–59. https://doi.org/10/ghstcv.

Hoffmann, Friedrich. "Der Bahnhof der Berlin-Hamburger Eisenbahn in Berlin." *Zeitschrift für Bauwesen* 6 (1856): 487–96.

Steinle, Holger. *Ein Bahnhof auf dem Abstellgleis. Der Ehemalige Hamburger Bahnhof in Berlin und seine Geschichte*. Berlin: Silberstreif, 1983.

Stüler, Friedrich August, and Johann Heinrich Strack. "Entwurf zu einem Abfahrtsgebäude für Eisenbahnen oder einem Bahnhof." *Architektonisches Album. Redigirt vom Architekten-Verein zu Berlin durch Stüler, Knoblauch, Salzenberg, Strack, Runge* 2 (1838): 4–8, XI, XII.

8 Görlitzer & Küstriner Bahnhof

Demps, Laurenz. *Der Schlesische Bahnhof in Berlin. Ein Kapitel preußischer Eisenbahn-Geschichte*. Berlin: TranzPress, 1991.

Galli, Emil. *Görlitzer Bahnhof – Görlitzer Park*. Berlin: Verein Görlitzer Park e.V., 1995.

Maier, Helmut. *Berlin Görlitzer Bahnhof*. Berlin: Privatdruck, 2016.

Nickel, Angela. "Ein Architekt im Übergang. August Orth (1828–1901)." *Berliner Monatsschrift* 3 (1996): 36–42.

Orth, August. "Der Bahnhof der Berlin-Görlitzer Eisenbahn zu Berlin." *Zeitschrift für Bauwesen* 22 (1872): 547–52.
Uebel, Lothar. *Eisenbahner, Artisten und Zeitungsmacher. Zur Geschichte des ehemaligen Küstriner Bahnhofs*. Berlin: Grundstücksgesellschaft Franz-Mehring-Platz, 2011.
Winkler, Dirk. "Mittendrin und Rundherum." *Die Eisenbahn in Berlin*. Eisenbahn-Kurier Special 133 (2019): 10–29.

9 Lehrter Bahnhof

Anon. "Die Einführung der Berlin–Lehrter Eisenbahn in den Stadtbezirk Berlin und die Berliner Bahnhofsanlagen derselben." *Deutsche Bauzeitung* 5, no. 27/39 (1871): 212–14, 305–9.
Anon. "Das Projekt der Umwandlung des Lehrter PersonenBahnhofs in Berlin zu einem Ausstellungs-Gebäude." *Deutsche Bauzeitung* 16, no. 41 (1884): 242.
Architektenverein zu Berlin, and Vereinigung Berliner Architekten, eds. *Berlin und seine Bauten*. 3 vols. Berlin: Wilhelm Ernst & Sohn, 1896.
Becker, Tobias, Anna Littmann, and Johanna Niedbalski. *Die tausend Freunde der Metropole. Vol. 6. Kulturgeschichte der Moderne*. Bielefeld: Transcript, 2011.
Berlin und seine Eisenbahnen 1846 – 1896. Im Auftrage des Verein Deutscher Eisenbahnverwaltungen, Verband der Preussischen Eisenbahnen, Königlich Preussischer Minister der Oeffentlichen Arbeiten. Two volumes. Berlin & Heidelberg: Springer, 1896.
Bley, Peter. *150 Jahre Eisenbahn Berlin – Hamburg*. Berlin: Alba, 1996.
Brinkmann, Tobias. "Strangers in the City: Transmigration from Eastern Europe and Its Impact on Berlin and Hamburg 1880–1914." *Journal of Migration History* 2, no. 2 (September 30, 2016): 223–46. https://doi.org/10/gkr866.
Krings, Ulrich. *Deutsche Großstadt-Bahnhöfe des Historismus*. München: Prestel, 1985.
Romers, H. *De spoorwegarchitectuur in Nederland, 1841–1938*. Zutphen: De Walburg Pers, 1981.
Winkler, Dirk. "Bahnhof am Hafen – der Lehrter Bahnhof." *Die Eisenbahn in Berlin*, Eisenbahn-Kurier Special, 133 (2019): 54–61.
Zschocke, Helmut. *Geheimnisvolles Alsenviertel am Bundeskanzleramt*. Frankfurt am Main: PL Academic Research, 2017.

10 The Beginning of the End of the Line

"B." "Der Auswanderer-Bahnhof in Ruhleben bei Spandau." *Centralblatt der Bauverwaltung* XIII (1893): 142–44.
Bley, Peter. *Friedhofsbahn Wannsee–Stahnsdorf*. Berlin: B. Neddermeyer, 2022.
Bley, Peter. *Friedhofsbahn Wannsee–Stahnsdorf. Eine fast vergessene S-Bahn-Strecke im Südwesten von Berlin*. Berlin: GVE-Verlag, 2022.
Kiebert, Wolfgang. "Wannsee–Stahnsdorf. Eine S-Bahn-Strecke im Abseits." *Verkehrsgeschichtliche Blätter* 36, no. 6 (2009):
Nordhausen, Richard, and W. Zehme. "Der Auswanderer-Bahnhof in Ruhleben." *Die Gartenlaube 1895*, no. 9 (1895): 140–42.
Pierson, Kurt. *Die Königl. Preußische Militär-Eisenbahn*. Stuttgart: Motorbuch Verlag, 1979.

Roloff, Carl. "Die Friedhofsbahn Wannsee–Stahnsdorf." *Zentralblatt der Bauverwaltung* 34, no. 83/ 84 (1914):

Rothe, Michael. "Bahnhof in die Freiheit: Der Auswandererbahnhof." *Berliner Verkehrsblätter* 59, no. 7 (2012): 124–28.

Schulz, Karin. "Der Auswandererbahnhof Ruhleben – Nadelöhr zum Westen." In *Die Reise nach Berlin*, 237–41. Berlin: Siedler, 1987.

Winkler, Dirk. "Bahnhof am Hafen – der Lehrter Bahnhof." *Die Eisenbahn in Berlin, Eisenbahn-Kurier Special* 133 (2019): 54–61.

Zureck, Hansjörg F. "100 Jahre Friedhofsbahn." *Berliner Verkehrsblätter*, no. 6 (2013):

11 Plans! Plans! Plans!

Bernhard, Patrick. "Metropolen auf Achse. Städtebau und Großstadtgesellschaften Rom und Berlins im faschistischen Bündnis 1936–1943." In *Berlin Im Nationalsozialismus: Politik Und Gesellschaft 1933–1945*, edited by Rüdiger Schaarschmidt Hachtmann, Thomas, and Winfried Süß, 132–57. Stuttgart: Wallstein, 2012.

Eiselen, Fritz. "Die Lösung der Verkehrsfragen im Wettbewerb Groß-Berlin." *Deutsche Bauzeitung*, no. 50–59 (1910): 385–392, 401.

Gröninck, E. *Berlin und seinen zukünftigen Central-Bahnhofs- und Central-Hafen-Anlagen*. Berlin: Polytechnische Buchhandlung A. Seydel, 1901.

Hofmann, Albert. "Groß-Berlin, sein Verhältnis zur modernen Großstadtbewegung und der Wettbewerb zur Erlangung eines Grundplanes für die städtebauliche Entwicklung Berlins und seiner Vororte im zwanzigsten Jahrhundert." *Deutsche Bauzeitung* 44 (1910): 169–181, 197, 213, 233, 261, 281, 311, 325.

Koll and Helm. "Der Verkehr in Gross-Berlin." *Verkehrstechnische Woche* 5, no. 11, 14, 21, 26, 28, 30 (1911).

Kuhlmann, Bernd. *Eisenbahn-Großenwahn in Berlin. Die Planungen von 1933 bis 1945 und deren Realisierung*. Berlin: GVE, 1996.

Reichhardt, Hans J., and Wolfgang Schäche. *Von Berlin nach Germania. Über die Zerstörungen der Reichshauptstadt durch Albert Speers Neugestaltungsplanen*. Berlin: Transit, 1985.

Roos, Harald. "Berlins Fern- und Nahverkehr. Vorschlag zu einer grundlegenden Neuordnung." *Das neue Berlin* 1, no. 7 (1929): 142–43.

Schützler, Heiko. "Monsalvat an der Spree: nationalsozialistische Planungen in Berlin, Welthauptstadt Germania." *Berlinische Monatsschrift* 9, no. 9 (2000): 27–35.

Tubbesing, Markus. *Der Wettbewerb Gross-Berlin 1910. Die Entstehung einer modernen Disziplin Städtebau*. Tübingen & Berlin: Wasmuth, 2018.

Winkler, Dirk. "Bahnhof am Hafen – der Lehrter Bahnhof." *Die Eisenbahn in Berlin, Eisenbahn-Kurier Special* 133 (2019): 54–61.

Table of Figures

https://doi.org/10.1515/9783111381879-013

Index

https://doi.org/10.1515/9783111381879-014

www.ingramcontent.com/pod-product-compliance
Lightning Source LLC
Chambersburg PA
CBHW070407100426
42812CB00005B/1667

.